# Mentals 6

Alan McSeveny Rachel McSeveny Diane McSeveny-Foster

**Pearson Australia**
(a division of Pearson Australia Group Pty Ltd)
459–471 Church St, Level 1, Building B, Richmond, Victoria, 3121
PO Box 23360, Melbourne, Victoria 8012
www.pearson.com.au


First published 2024 by Pearson Australia
2028 2027 2026 2025
10 9 8 7 6 5 4 3 2 1



Publishers: Sophie Matta and Kerry Nagle
Project Manager: Michelle Thomas
Production Editor: Laura Rentsch
Editor: Katie Millar
Designer: Anne Donald
Proofreader: Ann M. Philpott
Rights & Permissions Editor: Alice McBroom
Cover art: Michael Barter
Illustrator: Michael Barter
Publishing Services: Jit-Pin Chong
Printed in Australia by Pegasus Media + Logistics

ISBN 978 0 6557 0886 5
Pearson Australia Group Pty Ltd ABN 40 004 245 943

**Attributions**
We would like to thank the following for permission to reproduce copyright material.

**Shutterstock:** Ilya Akinshin, p. 17 (tv screen); Vovan, p. 53 (laptop).

**Acknowledgement of Country**
Pearson respects and honours Aboriginal and Torres Strait Islander Elders past, present and future. We acknowledge the stories, traditions and living cultures of the Traditional Custodians of the lands on which our company is located and where we conduct our business. Pearson is committed to honouring Australian Aboriginal and Torres Strait Islander peoples' unique cultural and spiritual relationships to the land, waters and seas and their rich contribution to society.

Aboriginal and Torres Strait Islander peoples are advised that this text may contain images, voices and names of deceased persons.

# Introduction

## Using the Mentals Books

This book reviews content from the Signpost Student Book. It is used most effectively when it aligns with the suggested program in the Student Book contents.

Each unit of the Mentals Book is programmed to review Student Book content for the previous two weeks. (The Suggested Program overview can be found in the Teacher Resource.) For example, Unit 15 of the Mentals Book can be set as homework to review weeks 13 and 14 of the Student Book while week 15 is being taught. Units 1 and 2 review work taught in the previous year.

## Mixed-topic questions

The units present questions in a mixed-topic format to encourage thorough understanding and continuous review.

## Graded questions

- Column 1: easier
- Columns 2 and 3: harder
- Column 4: Extension and Challenge

## Presentation

- Number facts are reinforced to encourage instant recall.
- Essential skills are explained.
- The Arithmetic card (page 5) is a useful teaching tool for practising basic number skills.
- ID cards (pages 6 to 9) review the mathematical terms students need to learn.
- Measurement benchmarks and Tables of number and measurement (pages 84 and 85) are provided so that students can learn important facts and estimate measurements effectively.

## Motivation

- There are two lizards hidden on each page for students to find.
- The header allows students to record their score.

### Extra activities

- Problem-solving **strategies** are introduced in a carefully planned sequence throughout the series.

- Important concepts from **Number and algebra** and **Measurement and geometry** are explored.

- **Measurement** concepts and activities are introduced and investigated.

- **Statistics and probability** concepts (Data and chance) are presented for revision and extension.

- A **tables** program for each of the four operations is included.
- It is important for students to learn addition and multiplication tables by heart.

# 6 Contents

## Teaching ideas using headers

| Unit | Content | Extra Activity |
|---|---|---|
| **1:1/2**<br>**1:3/4** | + 3, + 5<br>Personal measurements | + tables<br>Measure |
| **2:1/2**<br>**2:3/4** | − 2, − 4<br>Language | − tables<br>ID card D |
| **3:1/2**<br>**3:3/4** | × 8, × 5<br>Rounding money | × tables<br>Concept |
| **4:1/2**<br>**4:3/4** | × 2, × 4<br>+ 4, + 6 | × tables<br>+ tables |
| **5:1/2**<br>**5:3/4** | Percentages<br>Equivalent fractions | Concept<br>Concept |
| **6:1/2**<br>**6:3/4** | Order of operations<br>Square numbers / Multiples | Concept<br>Concept |
| **7:1/2**<br>**7:3/4** | Problem solving<br>Reflections | Strategy time<br>Concept |
| **8:1/2**<br>**8:3/4** | Square numbers<br>Order of operations | Concept<br>Concept |
| **9:1/2**<br>**9:3/4** | Language<br>÷ 2, ÷ 4 | ID card B<br>÷ tables |
| **10:1/2**<br>**10:3/4** | Language<br>−13, −17 | ID card B<br>− tables |
| **11:1/2**<br>**11:3/4** | × 3 × 6<br>Multiplication | × tables<br>× tables |
| **12:1/2**<br>**12:3/4** | ÷ 5, ÷ 10<br>Scale drawing | ÷ tables<br>Concept |
| **13:1/2**<br>**13:3/4** | ÷ 3, ÷ 6<br>× 6 × 9 | ÷ tables<br>× tables |
| **14:1/2**<br>**14:3/4** | Language<br>Language | ID card C<br>ID card C |
| **15:1/2**<br>**15:3/4** | Averages<br>× 7, × 8 | Concept<br>× tables |
| **16:1/2**<br>**16:3/4** | ÷ 9<br>Profit and loss | ÷ tables<br>Concept |
| **17:1/2**<br>**17:3/4** | Money<br>Chance | Strategy time<br>Chance |
| **18:1/2**<br>**18:3/4** | ÷ 7, ÷ 8<br>Money | ÷ tables<br>Strategy time |
| **19:1/2**<br>**19:3/4** | Problem solving<br>Problem solving | Strategy time<br>Strategy time |

| Unit | Content | Extra Activity |
|---|---|---|
| **20:1/2**<br>**20:3/4** | – 9, – 5<br>+ 7, + 9 | – tables<br>+ tables |
| **21:1/2**<br>**21:3/4** | Language<br>Crossnumber puzzle | ID card C<br>Concept |
| **22:1/2**<br>**22:3/4** | Magic squares<br>Crossnumber puzzle | Concept<br>Concept |
| **23:1/2**<br>**23:3/4** | – 3, – 5 – 9<br>Converting distances | – tables<br>Measure |
| **24:1/2**<br>**24:3/4** | Problem solving<br>Problem solving | Strategy time<br>Strategy time |
| **25:1/2**<br>**25:3/4** | Estimating measurements<br>Factors | Measure<br>Concept |
| **26:1/2**<br>**26:3/4** | Fractions (subtraction)<br>Fractions (subtraction) | Concept<br>Concept |
| **27:1/2**<br>**27:3/4** | Fractions to decimals<br>× 8, × 6 | Concept<br>× tables |
| **28:1/2**<br>**28:3/4** | Language<br>Problem solving | ID card A<br>Strategy time |
| **29:1/2**<br>**29:3/4** | Average speed<br>Problem solving | Measure<br>Strategy time |
| **30:1/2**<br>**30:3/4** | Problem solving<br>Codes | Strategy time<br>Concept |
| **31:1/2**<br>**31:3/4** | Order of operations<br>Scale drawing | Concept<br>Concept |
| **32:1/2**<br>**32:3/4** | × 6, × 7, × 8<br>÷ 4 | × tables<br>÷ tables |
| **33:1/2**<br>**33:3/4** | − 6, − 8<br>Roman numerals | – tables<br>Concept |
| **34:1/2**<br>**34:3/4** | Scale drawing<br>Coordinates | Concept<br>Concept |
| **35:1/2**<br>**35:3/4** | Codes<br>Factors | Concept<br>Concept |
| **36:1/2**<br>**36:3/4** | Tally<br>Divisibility | Chance<br>Concept |
| **37:1/2**<br>**37:3/4** | Coordinates with 4 quadrants<br>Personal measurements | Concept<br>Measure |
| **Answers** | These can be found in the middle of this book on pages A1 to A16. | |

## Arithmetic card

| | A | B | C | D | E | F | G | H | I | J | K | L | M |
|---|---|---|---|---|---|---|---|---|---|---|---|---|---|
| **1** | 20 | 9 | 110 | 20 | 6 | 11 | $100 | $\frac{67}{100}$ | 0·52 | 97% | $\frac{56}{100}$ | 20 | 67 |
| **2** | 16 | 1 | 170 | 50 | 14 | 21 | $32 | $\frac{3}{10}$ | 0·2 | 52% | $\frac{8}{100}$ | 50 | 62 |
| **3** | 12 | 6 | 200 | 10 | 2 | 27 | $48 | $\frac{75}{100}$ | 0·75 | 39% | $\frac{27}{100}$ | 10 | 50 |
| **4** | 18 | 2 | 130 | 70 | 18 | 25 | $90 | $\frac{46}{100}$ | 0·39 | 74% | $\frac{15}{100}$ | 70 | 41 |
| **5** | 13 | 8 | 160 | 100 | 10 | 13 | $117 | $\frac{5}{100}$ | 0·4 | 63% | $\frac{88}{100}$ | 100 | 33 |
| **6** | 15 | 5 | 180 | 30 | 16 | 17 | $150 | $\frac{9}{10}$ | 0·98 | 6% | $\frac{42}{100}$ | 30 | 26 |
| **7** | 19 | 3 | 150 | 80 | 4 | 29 | $76 | $\frac{14}{100}$ | 0·67 | 18% | $\frac{74}{100}$ | 80 | 19 |
| **8** | 14 | 7 | 190 | 60 | 12 | 15 | $28 | $\frac{83}{100}$ | 0·5 | 45% | $\frac{66}{100}$ | 60 | 14 |
| **9** | 11 | 4 | 140 | 90 | 20 | 23 | $55 | $\frac{6}{10}$ | 0·13 | 87% | $\frac{39}{100}$ | 90 | 7 |
| **10** | 17 | 10 | 120 | 40 | 8 | 19 | $85 | $\frac{1}{100}$ | 0·8 | 21% | $\frac{95}{100}$ | 40 | 3 |

### How to use this card

If students were told to 'subtract B from C', they would write:

❶ 110 – 9 = 101 ❷ 170 – 1 = 169 ❸ 200 – 6 = 194 ❹ 130 – 2 = 128 ❺ 160 – 8 = 152
❻ 180 – 5 = 175 ❼ 150 – 3 = 147 ❽ 190 – 7 = 183 ❾ 140 – 4 = 136 ❿ 120 – 10 = 110

Other instructions might be:

- **What is left from $200 if I spend the amount in column G?**
- **What multiplied by 10 gives column C?**
- **Subtract column B from L.**
- **Write column J as a decimal.**
- **Multiply column B by 4.**
- **Add columns A and C.**
- **Add columns H and K.**
- **Square column B.**
- **Halve column E.**

The applications of this card are endless.

# ID card A

Do not write on this card.

| 1 | 2 | 3 | 4 | 5 |
|---|---|---|---|---|
| **mm** stands for m______. | **cm** stands for c______. | **m** stands for m______. | **km** stands for k______. | 0 1 2 3 4 / 1 cm = 1 m s______ |

| 6 | 7 | 8 | 9 | 10 |
|---|---|---|---|---|
| **mg** stands for m______. | **g** stands for g______. | **kg** stands for k______. | **t** stands for t______. | **mL** stands for m______. |

| 11 | 12 | 13 | 14 | 15 |
|---|---|---|---|---|
| **L** stands for l______. | **kL** stands for k______. | **ML** stands for M______. | **cm²** stands for s______ c______. | **m²** stands for s______ m______. |

| 16 | 17 | 18 | 19 | 20 |
|---|---|---|---|---|
| 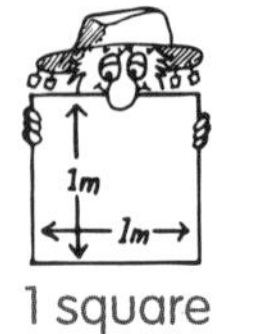  1 square m______ | **ha** stands for h______. | 100 m, 1 ha, 100 m — 1 hectare = ______ m² | **km²** stands for s______ k______. | **cm³** stands for c______ c______. |

| 21 | 22 | 23 | 24 | 25 |
|---|---|---|---|---|
| **1 cm³** has the same volume as ______ mL of liquid. | **m³** stands for c______ m______. | **s** stands for s______. | **min** stands for m______. | **h** stands for h______. |

| 26 | 27 | 28 | 29 | 30 |
|---|---|---|---|---|
| **am** means b______ n______. | **pm** means a______ n______. | **05:40** in digital time is ____:____ ____. | **7:30 pm** in 24-hour time is ____:____. | **°C** stands for d______ C______. |

See page A1 for answers.



 ISBN 978 0 6557 0886 5

# ID card B

Do not write on this card.

**1** …, −3, −2, −1, 0, 1, 2, 3, … These are called i_________.

**2** A counting number that has only 2 factors, itself and 1 is a p_________ number.

**3** A counting number that has more than 2 factors is a c_________ number.

**4** 7, 14 21, … are multiples of _________.

**5** The factors of 8 are 1, 8, _____ and _____.

**6** 35, 40 and 45 are all divisible by _____.

**7** $5\overline{)47}$ = 9 *r* 2 'r' means r_________.

**8** 1, 4, 9, 16, 25, … are the s_________ numbers.

**9** 208 650 has 6 d_________.

**10** 4·395
____ ones
____ tenths
____ hundredths
____ thousandths

**11** 19% means 19 out of _________.

**12**

v_______ line

**13** h_______ line

**14** This part of a line is an i_________.

**15**

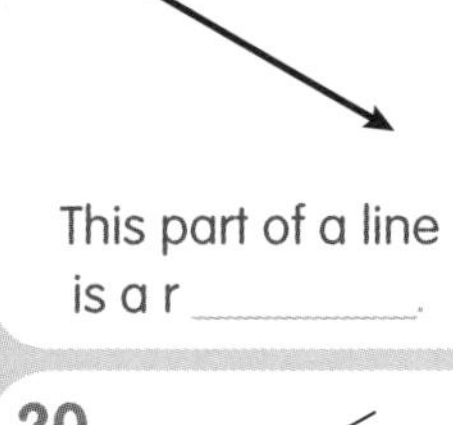

This part of a line is a r_________.

**16**

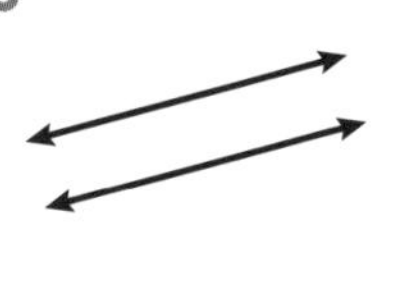

p_________ lines

**17** p_________ lines

**18**

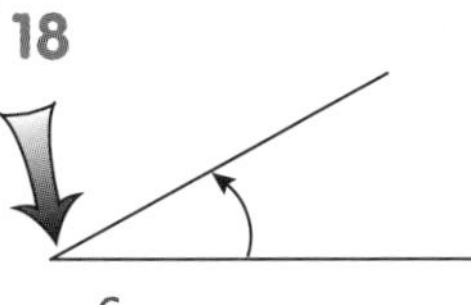

c_________ (or vertex of an angle)

**19** a_________ of an angle

**20** a_________ angle

**21**

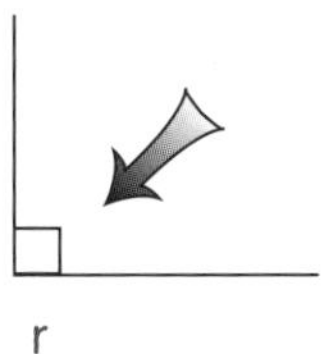

r_________ angle

**22** o_________ angle

**23** s_________ angle

**24**

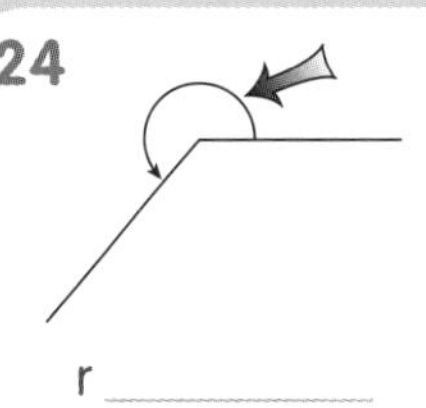

r_________ angle

**25**

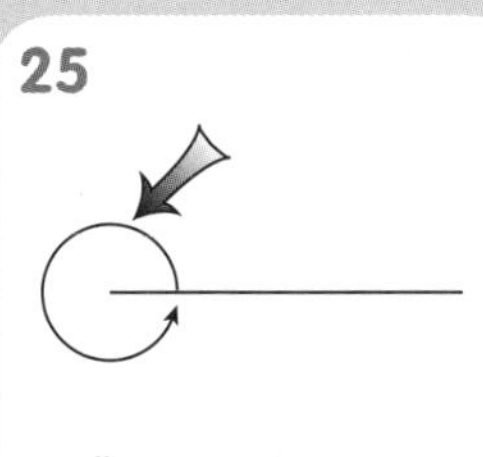

r_________.

**26** These angles are complementary.

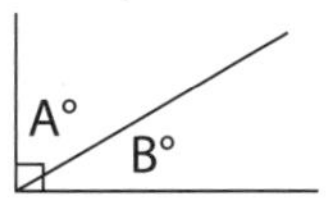


A° + B° = _________

**27** These angles are supplementary.

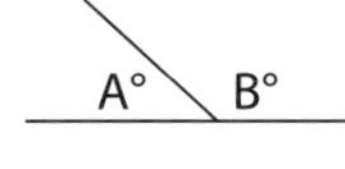


A° + B° = _________

**28** Angles at a point

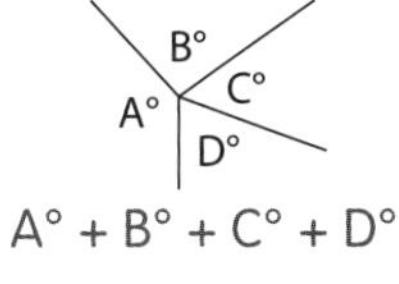


A° + B° + C° + D° = _________

**29** Vertically opposite angles

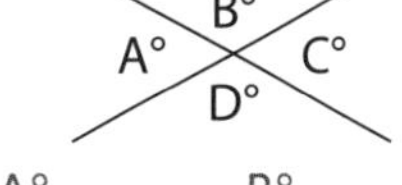


A° = _____, B° = _____

**30**

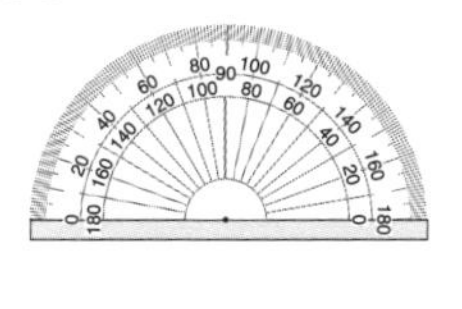

p_________

**See page A1 for answers.**

# ID card C

ID Card C

Do not write on this card.

**1**

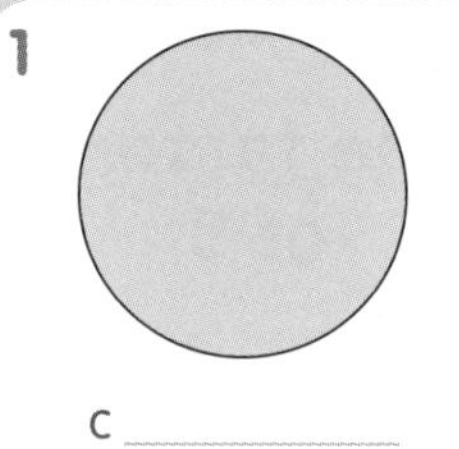

c ______

**2**

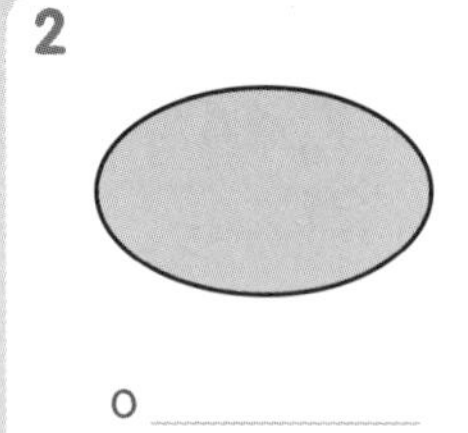

o ______

**3**

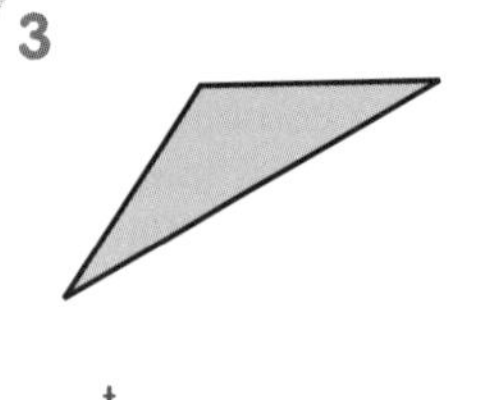

t ______

**4**

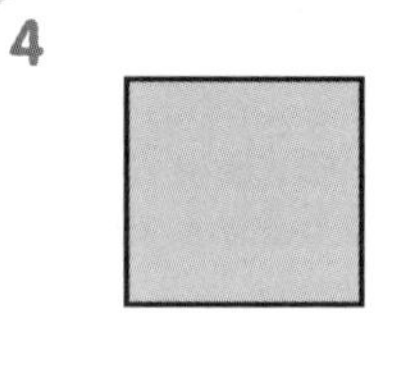

s ______

**5**

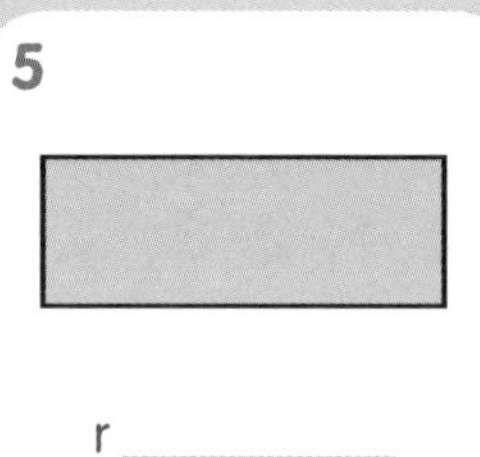

r ______

**6**

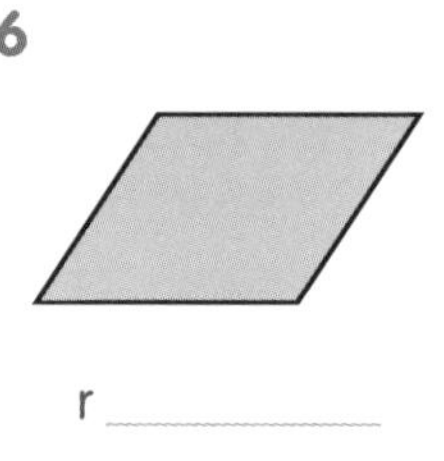

r ______

**7**

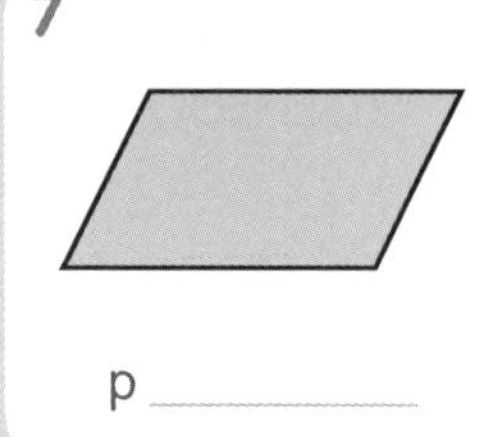

p ______

**8**

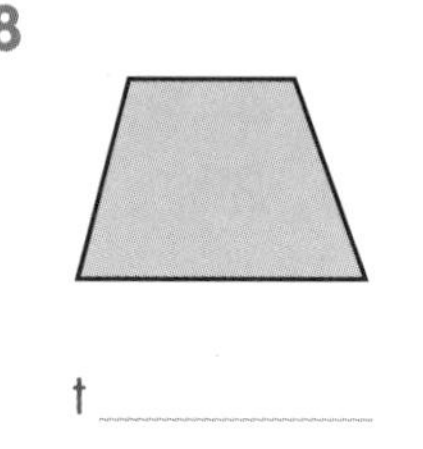

t ______

**9**

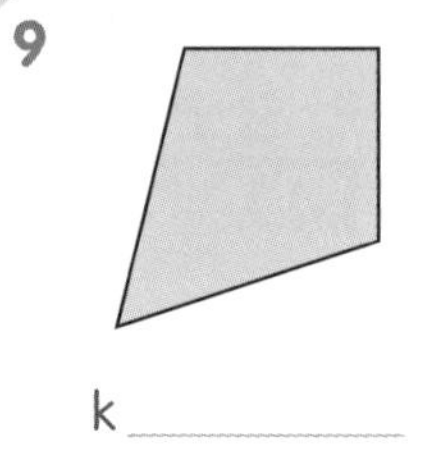

k ______

**10**

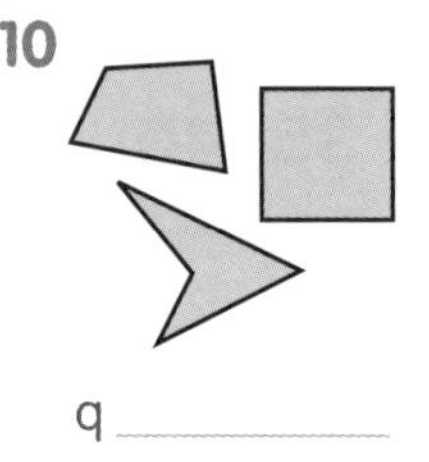

q ______

**11**

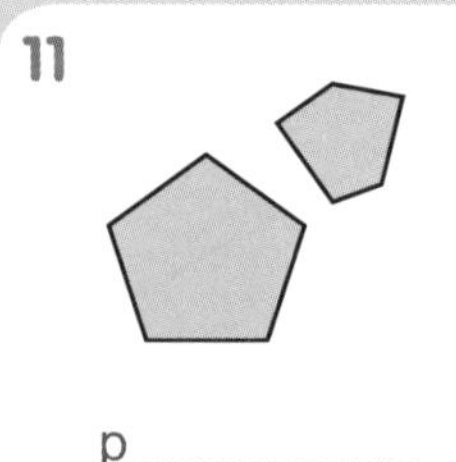

p ______

**12**

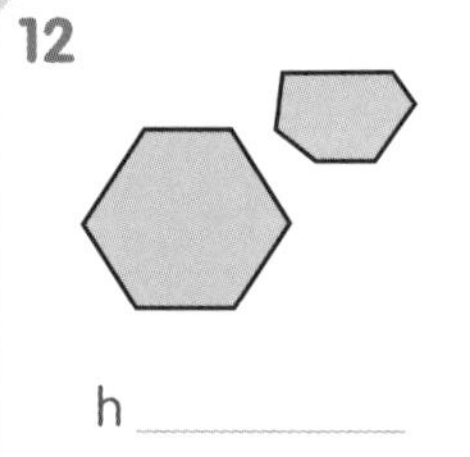

h ______

**13**

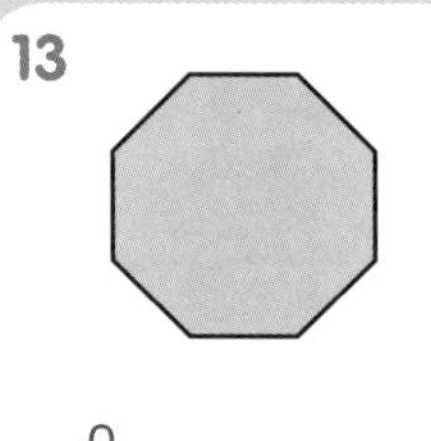

o ______

**14**

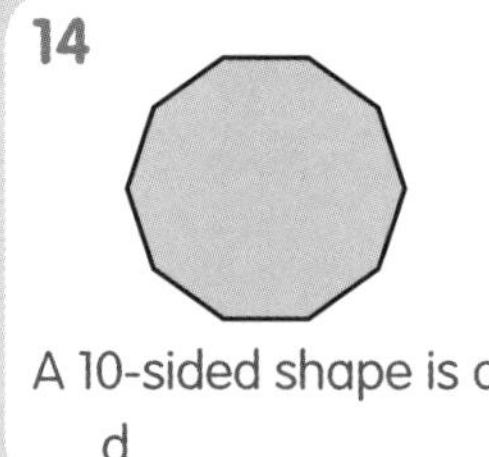

A 10-sided shape is a d ______.

**15**

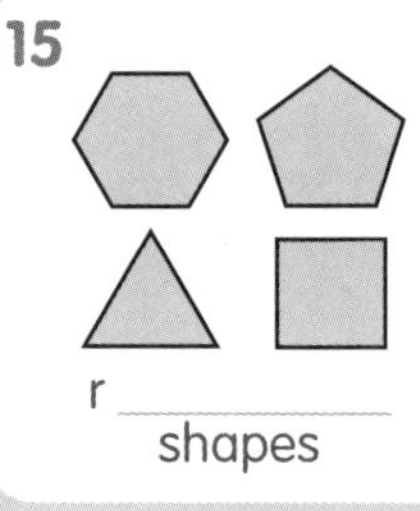

r ______ shapes

**16**

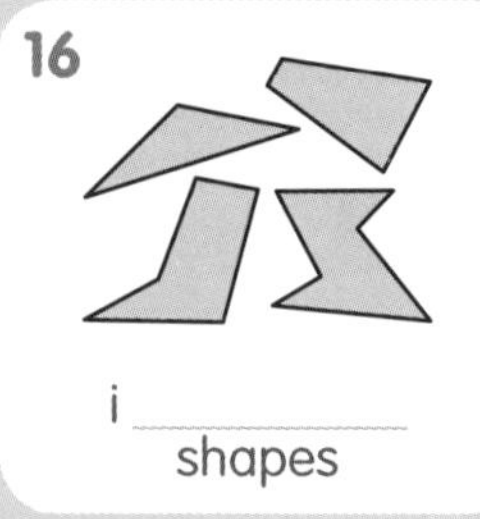

i ______ shapes

**17**

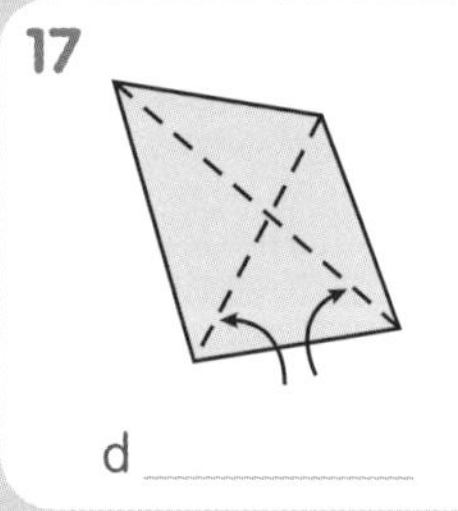

d ______

**18**

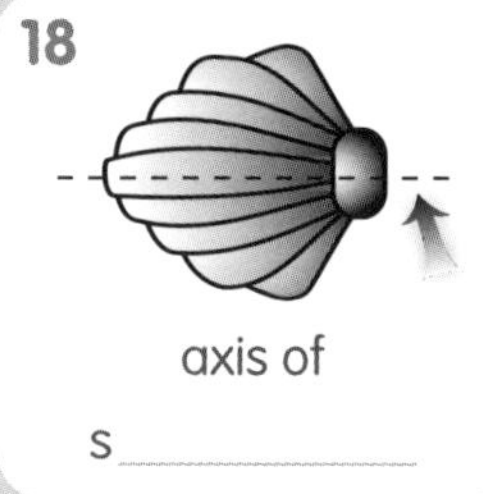

axis of s ______

**19**

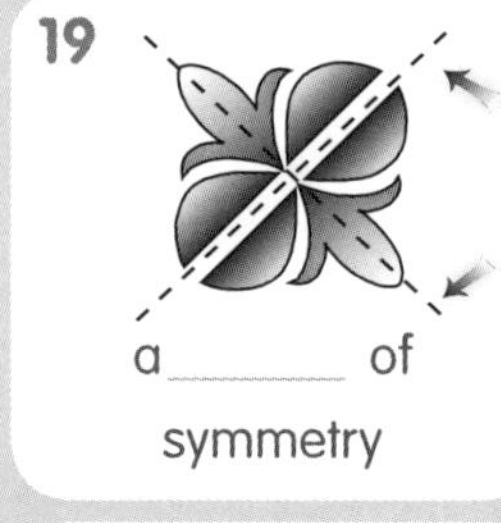

a ______ of symmetry

**20**

This has r ______ s ______.

**21**

r ______

**22**

t ______

**23**

r ______

**24**

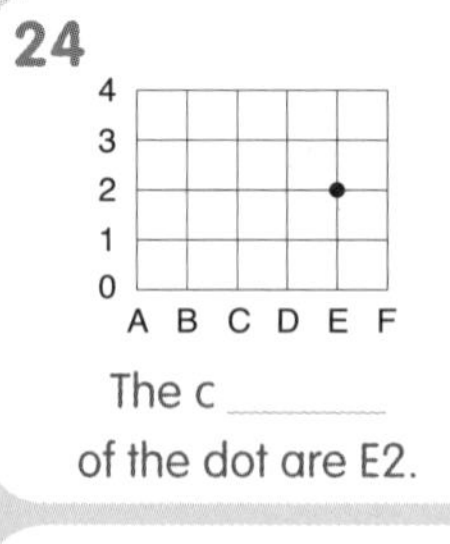


The c ______ of the dot are E2.

**25**

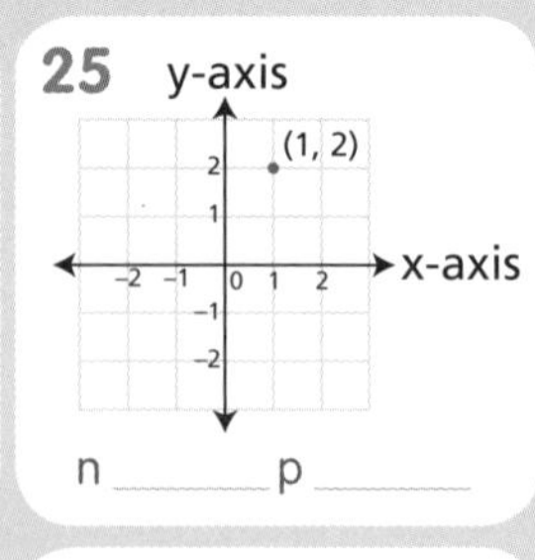


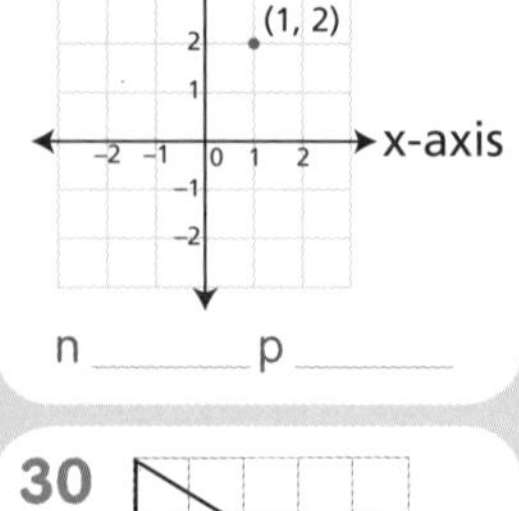

n ______ p ______

**26**

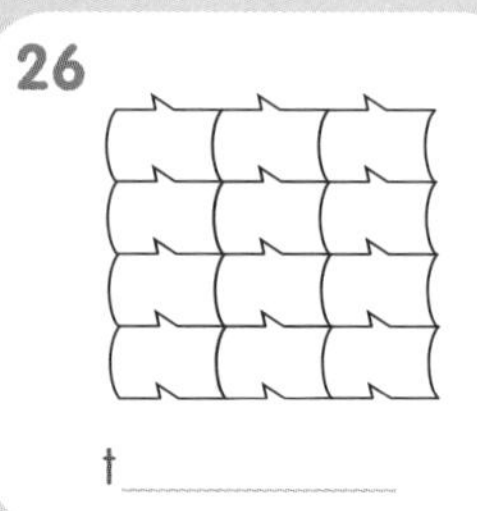

t ______

**27**

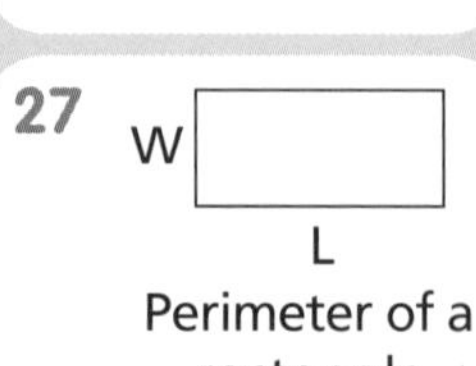

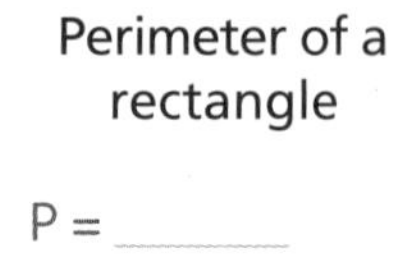

W

L

Perimeter of a rectangle

P = ______

**28**

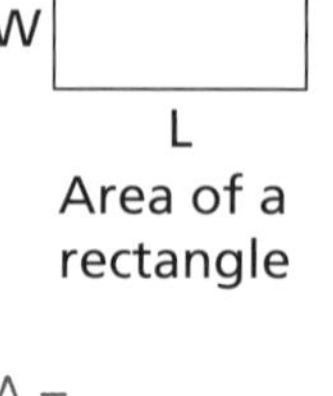

W

L

Area of a rectangle

A = ______

**29**

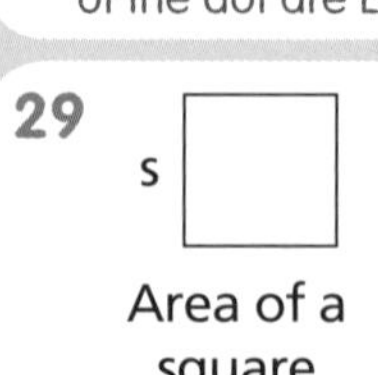

s

Area of a square

A = ______

**30**

The area of a triangle is half the area of a r ______.

See page A1 for answers.

# ID card D

Do not write on this card.

**1**

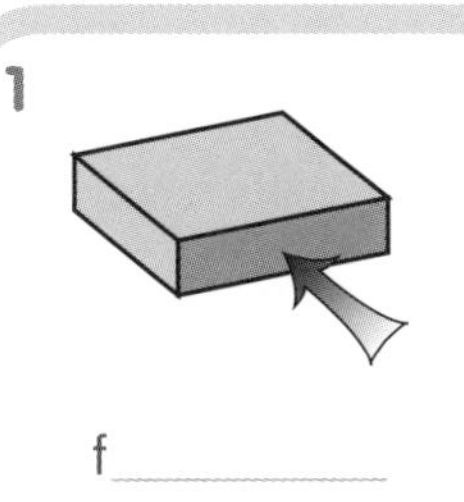

f

**2**

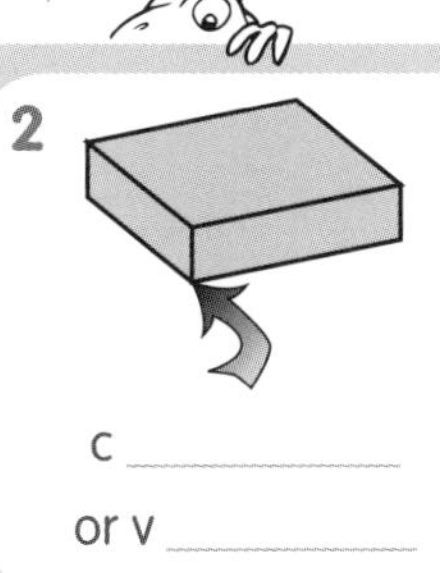

c
or v

**3**

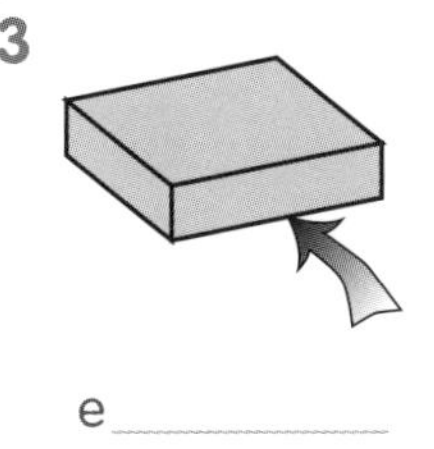

e

**4**

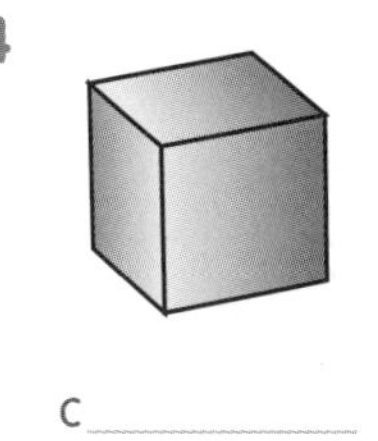

c

**5**

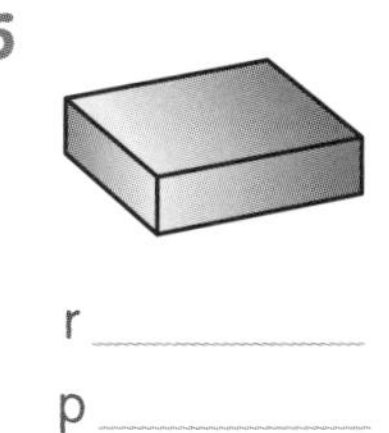

r
p

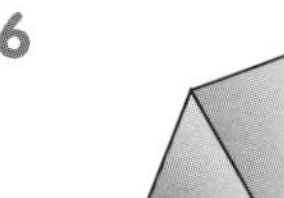

**6**

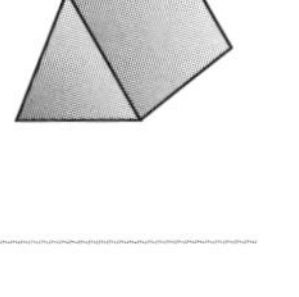

t
p

**7**

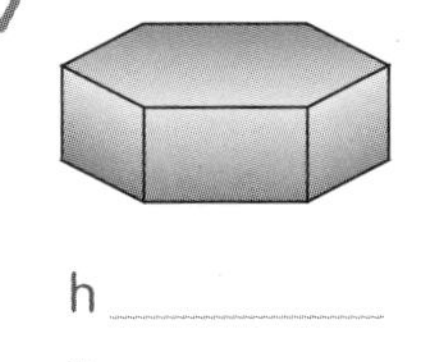

h
p

**8**

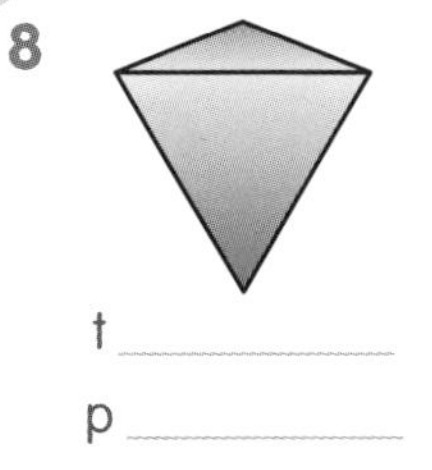

t
p

**9**

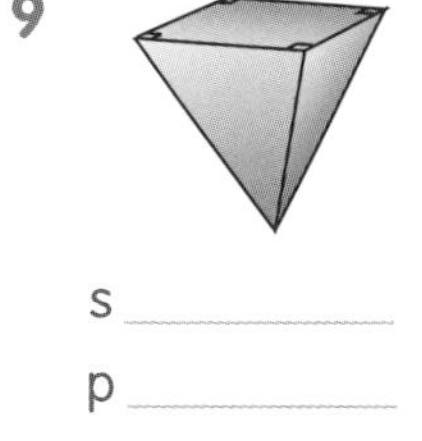

s
p

**10**

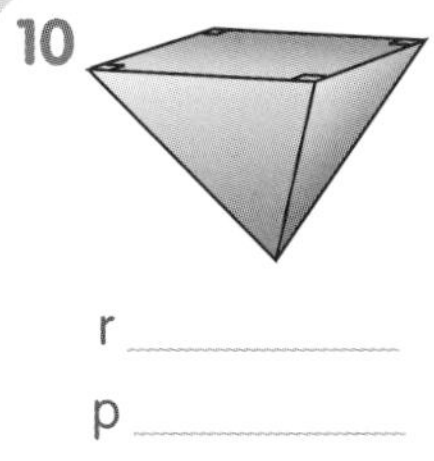

r
p

**11**

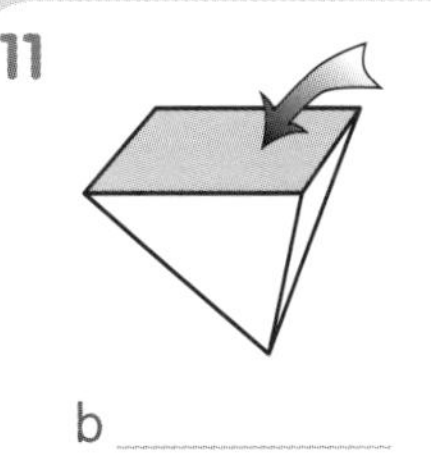

b

**12**

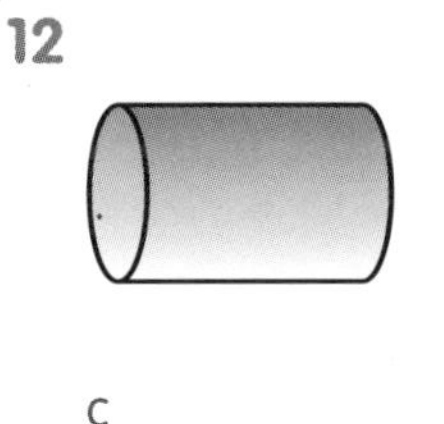

c

**13**

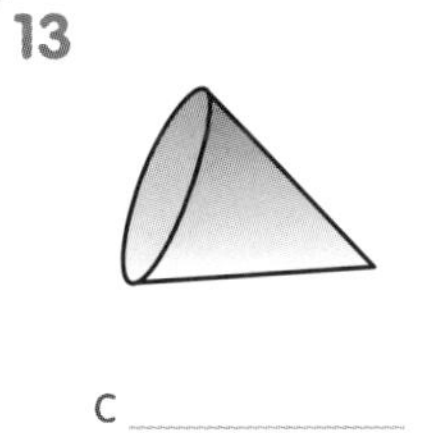

c

**14**

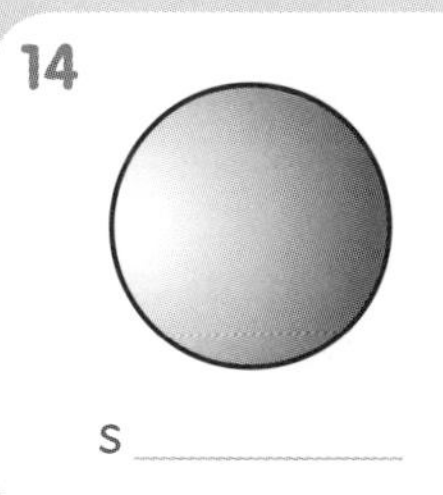

s

**15**

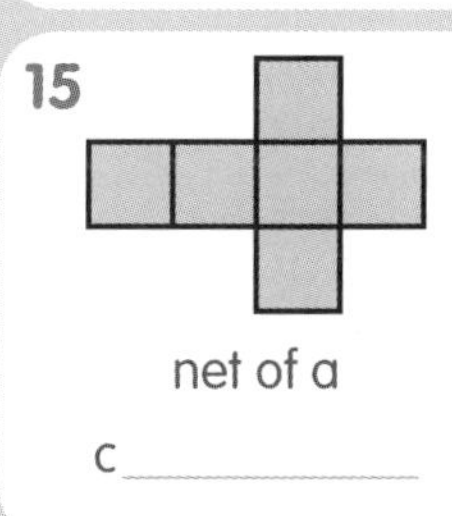

net of a
c

**16**

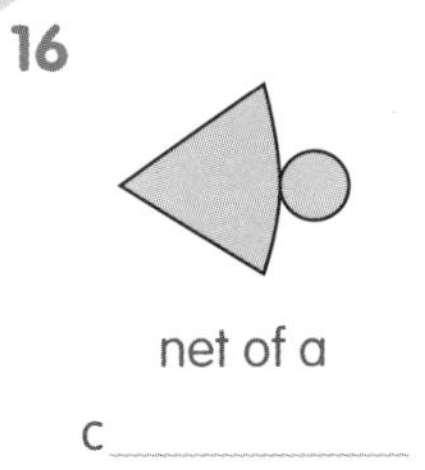

net of a
c

**17**

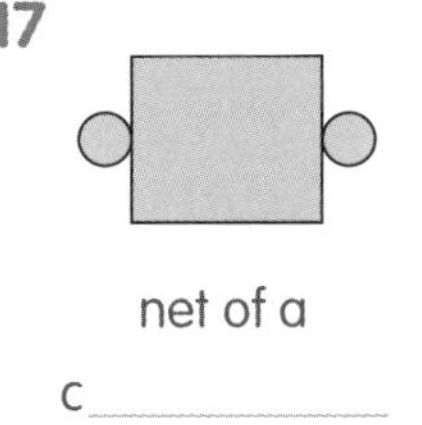

net of a
c

**18**

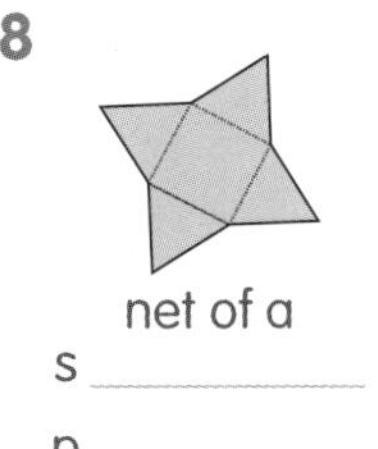

net of a
s
p

**19**

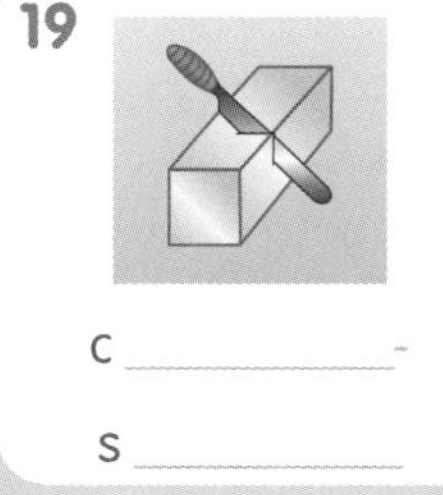

c -
s

**20**

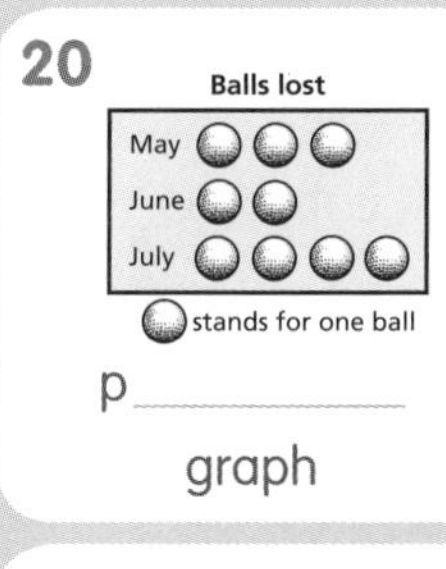


p
graph

**21**

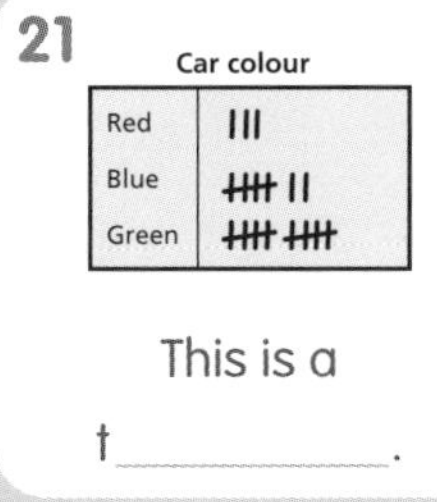


This is a
t .

**22**

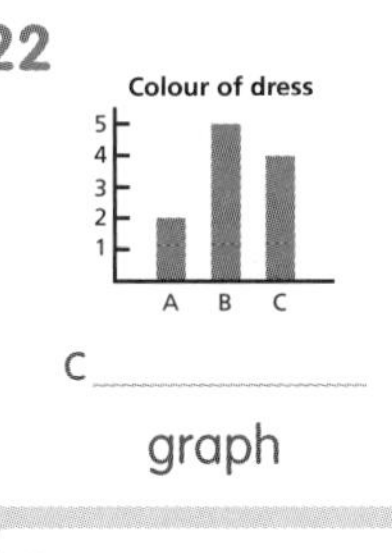


c
graph

**23**

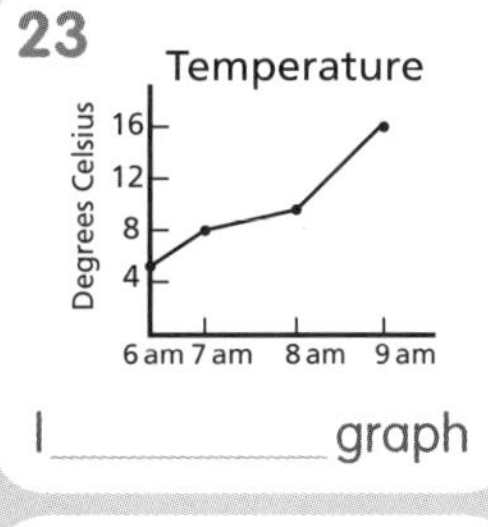


l graph

**24**

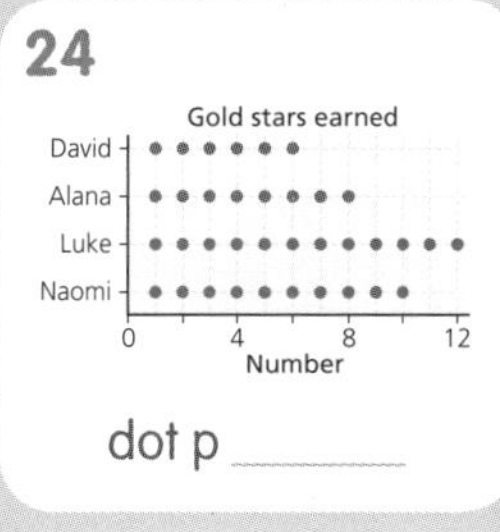


dot p

**25**

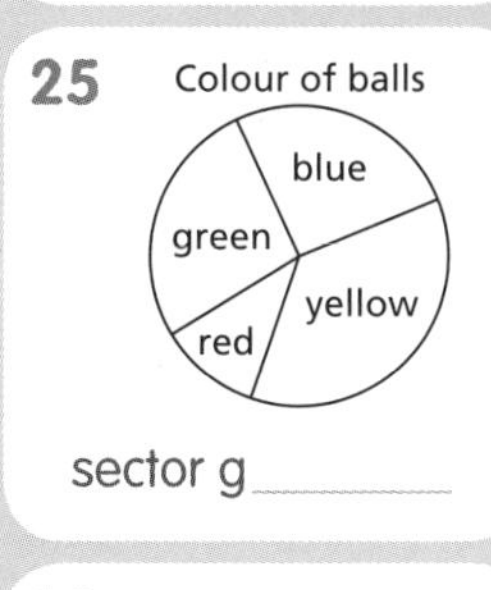


sector g

**26**

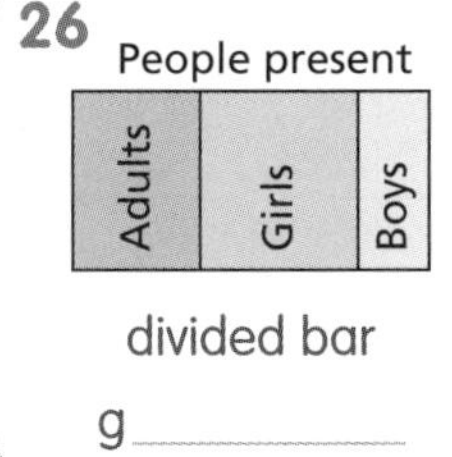


divided bar
g

**27**

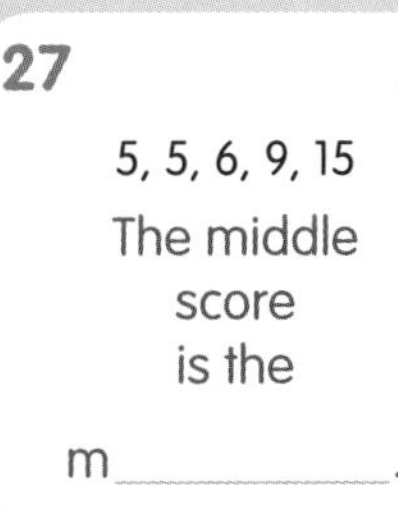

5, 5, 6, 9, 15
The middle score is the
m .

**28**

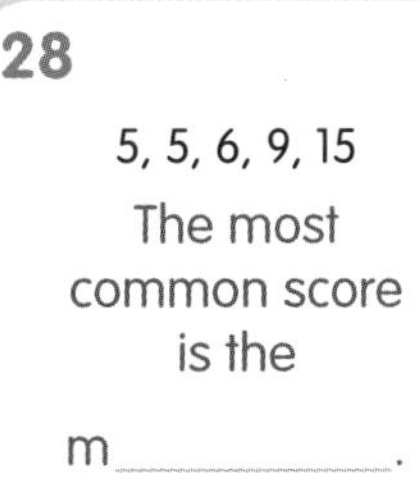

5, 5, 6, 9, 15
The most common score is the
m .

**29** 5, 5, 6, 9, 15

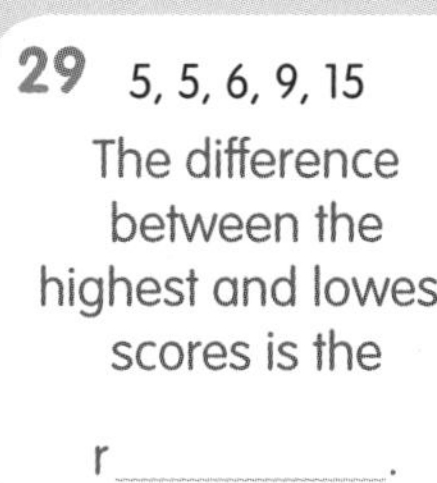

The difference between the highest and lowest scores is the
r .

**30** 5, 5, 6, 9, 15

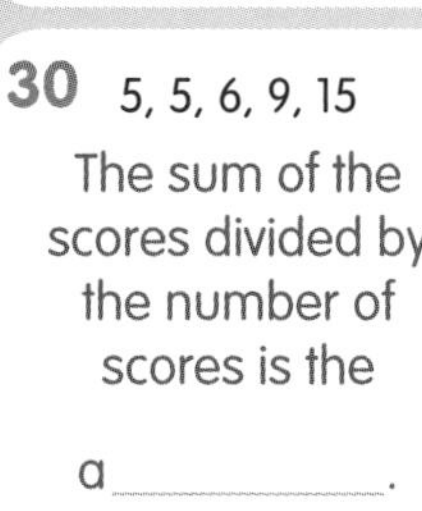

The sum of the scores divided by the number of scores is the
a .

**See page A1 for answers.**

## 1:1 out of 18

1. 23 + 31 ______
2. 3 × 3 ______
3. 6 × \$3 ______
4. 8 ÷ 2 ______
5. $\begin{array}{r} 5475 \\ +\ 2573 \\ \hline \end{array}$
6. Add 235 to 432. ______
7. Multiply 7 by 3. ______
8. 18 divided by 6. ______
9. 0·1 × 10 ______
10. $\begin{array}{r} 4253 \\ \times\quad 2 \\ \hline \end{array}$
11. 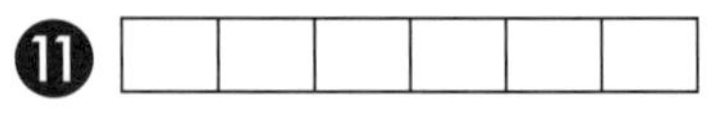 

    **a** $1 - \frac{5}{6} = \frac{\square}{\square}$ **b** $1 - \frac{2}{6} = \frac{\square}{\square}$
12. If 19·7 million is 19 700 000, what is:

    49·3 million? ______
13. $\frac{6}{8} - \frac{2}{8} = \frac{\square}{\square}$ or $\frac{\square}{\square}$
14. **a** 0·48 = ______% **b** 0·29 = ______%
15. **a** 20, 24, 28, ______, ______, ______, ______

    **b** 15, 18, 21, ______, ______, ______, ______

    **c** 22, 27, 32, ______, ______, ______, ______
16. A cube has ______ edges.
17. **a** 256 cm = ______ m ______ cm

    **b** 46 mm = ______ cm

    **c** 4 kg = ______ g

    **d** 4750 g = ______ kg
18. The total value of these notes.

    ______

See page 85.

## 1:2 out of 20

1. 64 ÷ 8 ______
2. 7 × \$6 ______
3. 648 − 97 ______
4. $\frac{1}{2}$ of 86 ______
5. $\begin{array}{r} 8353 \\ -\ 2647 \\ \hline \end{array}$
6. 3 squared. ______
7. Halve \$64. ______
8. 684 − 298 ______
9. 0·5 × 100 ______
10. $\begin{array}{r} 5308 \\ \times\quad 5 \\ \hline \end{array}$
11. Millilitres in $6\frac{1}{2}$ L. ______
12. Make the denominators equal, then subtract.

    $\frac{8}{12} - \frac{1}{3} = \frac{\square}{\square} - \frac{\square}{\square} = \frac{\square}{\square}$

13. **a** 0·54, 0·53, 0·52, ______, ______, ______

    **b** The rule for this pattern is ______.
14. 4 × 164 = (4 × ___) + (4 × ___) + (4 × ___)

    = ______ + ______ + ______

    = ______

    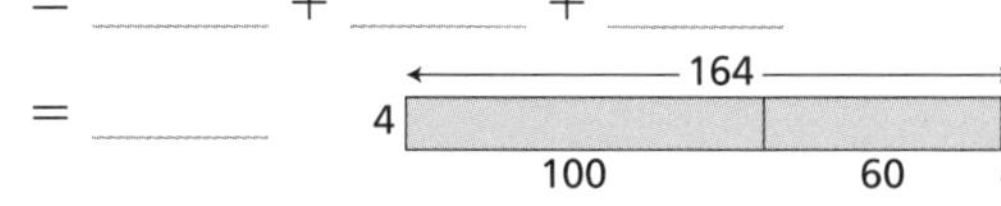
    
15. If apples cost 90c each,
    how much would 7 apples cost? ______
16. Find an estimate by rounding each number first.

    **a** 36 × 29 ______ **b** 64 × 17 ______

    **c** 53 × 31 ______ **d** 69 × 62 ______
17. $5\frac{1}{6}, 5\frac{2}{6}, 5\frac{3}{6}, \frac{\square}{\square}, \frac{\square}{\square}, \frac{\square}{\square}$
18. 5 hours = ______ minutes
19. Write the fraction equal to zero point nine. $\frac{\square}{\square}$
20. I scored 68 out of 100 in a test.
    What percentage did I get correct? ______

+ Tables

| 76 | 17 | 41 | 23 | + 3 | 30 | 55 | 52 | 44 | 39 | 28 |
|---|---|---|---|---|---|---|---|---|---|---|

| 76 | 17 | 41 | 23 | + 5 | 30 | 55 | 52 | 44 | 39 | 28 |
|---|---|---|---|---|---|---|---|---|---|---|

even + odd = ______

odd + odd = ______



 • *AUSTRALIAN SIGNPOST MATHS 6 MENTALS* • ISBN 978 0 6557 0886 5

## 1:3 ☐ out of 18

1. $8\overline{)864}$
2. $6\overline{)762}$
3. $5\overline{)835}$
4. 700 000 + 45 000 + 300 + 21 = ______
5. List the first 6 multiples of 7.

   ______, ______, ______, ______, ______, ______
6. Make the largest possible 8-digit number using 4, 2, 7, 4, 7, 9, 1 and 7. ______
7. In every space made by a row of 6 trees, we planted 8 flowers. How many flowers did we plant? ______
8. How many groups of 8 apples in 48? ______
9. Find the area of a rectangle with 4 m and 9 m sides. ______
10. 67 − 45 = 62 − ______
11. 500 ha = ______ square kilometres
12. What are the factors of 12?

    ______, ______, ______, ______, ______, ______
13. Colour 6 tenths red and 3 tenths blue. What fraction have you coloured? ______

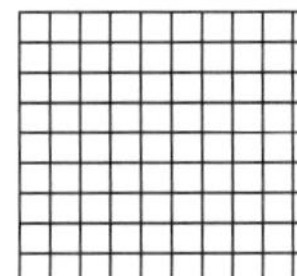

14. a 255 more than 121. ______

    b 243 more than 456. ______

    c 372 more than 627. ______
15. Which of 9, 21, 32 and 36 is both 'even' and 'a multiple of 3'? ______
16. a 185, 175, 165, ______, ______, ______

    b 16, 24, 32, ______, ______, ______, ______
17. 17 + 4 + 8 + 13 + 9 + 6 + 20 ______
18. 81 pens shared equally by 9. ______

## 1:4 Extension ☐ out of 8

1. January 1st is the 1st day of the year. What day is May 15th, 2023? ______
2. Fill in the boxes.

   a ☐☐ × 3 = ☐92

   b $3\overline{)☐☐☐}$ = 129
3. 3 × 3 × 90 ______
4. a 108 more than 52. ______

   b 108 more than 352. ______

   c 108 more than 35 352. ______
5. I must choose an item from each square. How many groups are possible? ______

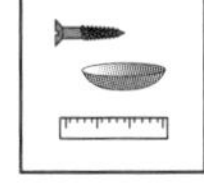

6. What fraction of this shape has been shaded? $\frac{☐}{☐}$

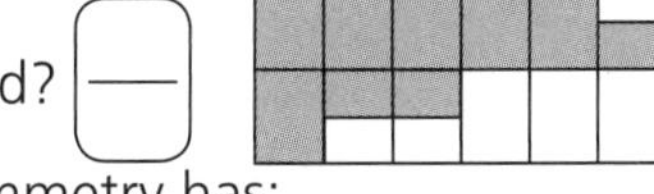

7. How many axes of symmetry has:

   a a regular heptagon? (7 sides) ______

   b a regular nonagon? (9 sides) ______
8. How many diagonals has a:

   a pentagon? ______ b hexagon? ______

### Challenge

*Write facts you know about the number 43·828.*

______

______

______

______

______

### Measure

Fill out this table about yourself, a relative or a friend.

**Name:** ______ **Date:** ______

| Age: ______ | Mass: ______ kg | Shoe size: ______ |
|---|---|---|
| Height: ______ cm | Waist: ______ cm | Neck size: ______ cm |

## 2:1 out of 19

1. $2 \times 3$ ______
2. $9 \times 2$ ______
3. $35 + 45$ ______
4. $38 + 62$ ______
5. $\begin{array}{r} 7456 \\ +\ 1456 \\ \hline \end{array}$
6. 99 more than 123. ______
7. Multiply 0·1 by 100. ______
8. 7 times \$4. ______
9. Divide 32 by 8. ______
10. $\begin{array}{r} 1432 \\ \times\ \ \ \ 5 \\ \hline \end{array}$
11. I scored 76 out of 100 in a test. How many more marks did I need to score 100? ______
12. Use am or pm to write 05:56. ______
13. A mango cost \$3.10. Circle the best estimate for the cost of 5 mangos.
    \$6 \$12 \$15 \$20 \$25
14. Estimate your height. ______
15. Write in order from largest to smallest.
    8 781 344, 8 768 367, 8 780 033
    ______
16. Round 46 354 to the nearest hundred. ______
17. How many tens can be taken from 354 838? ______
18. **a** 5000 mL = ______ L
    **b** 7·3 L = ______ mL
    **c** 7365 m = ______ km
    **d** 5 L 538 mL = ______ mL
19. **a** 18, 24, 30, ___, ___, ___, ___
    **b** 21, 28, 35, ___, ___, ___, ___

## 2:2 out of 19

1. $4 \times 8$ ______
2. $6 \times 5$ ______
3. $45 \div 9$ ______
4. $35 \div 7$ ______
5. $\begin{array}{r} 6475 \\ -\ 3659 \\ \hline \end{array}$
6. $246 - 137$ ______
7. $673 - 264$ ______
8. Multiply \$8 by 9. ______
9. 72 divided by 8. ______
10. $\begin{array}{r} 3751 \\ \times\ \ \ \ 3 \\ \hline \end{array}$
11. What is the probability, as a fraction, of tossing a 6 on a regular dice? $\frac{\square}{\square}$
12. Write 52 million. ______
13. What fraction of a dollar is 65c? $\frac{\square}{\square}$
14. Write 9 756 000 in words.
    ______
    ______
15. **a** On this square, shade 49 hundredths.
    **b** A batsman has scored 49 runs. How many more for a century? ______

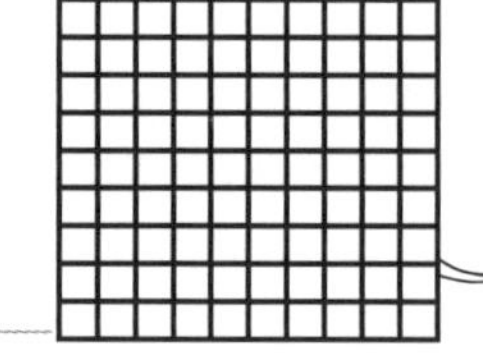

16. Write $2\frac{63}{100}$ as a decimal. ______
17. **a** $\frac{1}{4}$ of 12 ______
    **b** $\frac{3}{4}$ of 12 ______

18. **a** $2\frac{1}{2}$ m = ______ cm **b** $2\frac{1}{2}$ cm = ______ mm
19. The distance around each square is 16 metres. How far is it around the rectangle?

Tables

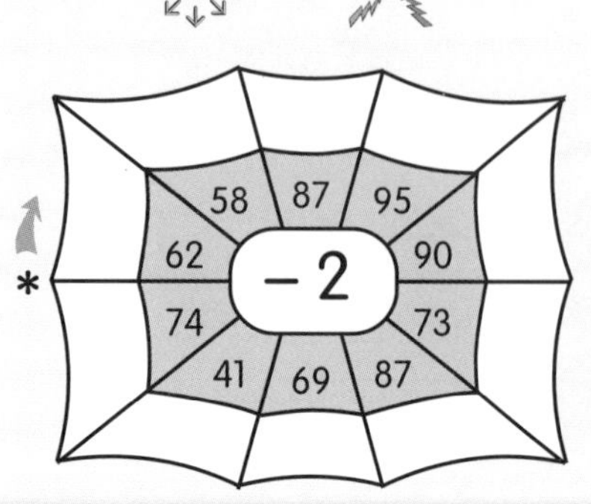


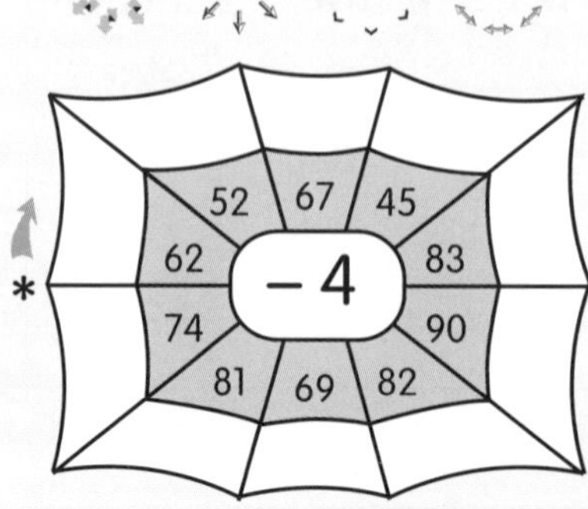

 ISBN 978 0 6557 0886 5

## 2:3 ☐ out of 12

1. $8\overline{)872}$ 2. $4\overline{)812}$ 3. $10\overline{)960}$
4. How many hundreds can be taken from 2 456 064? ______
5. Which is larger:
   a 6·09 or 4·87? ______
   b 17·25 or 19·3? ______
6. Our cricket team needed 260 runs to win the match. We scored only 123. How far short were we? ______

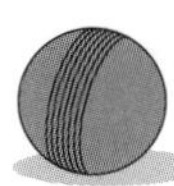

7. 9 o'clock, quarter past 12, half past 9, 3 o'clock. At which of the above times are the hour and minute hands at right angles?
   ______
8. What is the perimeter of the square that has an area of 16 $cm^2$? ______
9. What is the digital time 36 minutes after 15:27? ______
10. a What is the total capacity of these containers? ______

   b How many mL more would be needed to reach 2L? ______
11. 201·79 = ______ hundreds, ______ tens, ______ ones, ______ tenths, ______ hundredths
12. What fraction is:

   a shaded? $\frac{\square}{\square}$ b not shaded? $\frac{\square}{\square}$

## 2:4 Extension ☐ out of 6

1. How many axes of symmetry has:
   a a regular decagon? ______
   b a regular dodecagon? (12 sides) ______
2. This figure has hexagons of different size and shape. How many hexagons are there altogether? ______

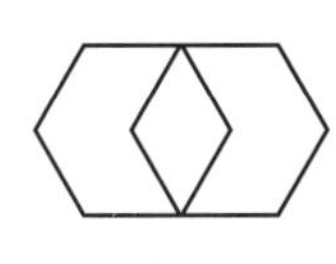

3. In how many different ways can you make 45 cents using only 5c, 10c and 20c coins? ______
4. A plane ticket to Armidale costs $174. What is the cost of:

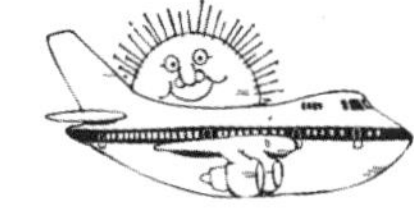

   a 5 tickets?
   b 6 tickets?
5. a 435 + 297 = ______ + 300 = ______
   b 824 − 392 = ______ − 390 = ______
   c 725 − 409 = ______ − 410 = ______
6. If the length, breadth and height of this model are all multiplied by 3, how many cubes will be in the new model's:

   a top view? ______
   b total volume? ______

**Challenge**

*Write number sentences that are equal to 36.*

______
______
______
______
______

Turn to ID card D on page 9.
Give the answers for these numbers.

(1) ______ (2) ______
(3) ______ (12) ______
(13) ______ (14) ______
(15) net of a ______ (16) net of a ______
(17) net of a ______ (18) net of a ______

A corner is also called a vertex.

## 3:1 out of 14

1. $4 \times 5$ ______
2. $5 \times 3$ ______
3. $6 \div 2$ ______
4. $12 \div 3$ ______
5. $\begin{array}{r} 6576 \\ -\ 3657 \\ \hline \end{array}$
6. Halve 282. ______
7. Double 432. ______
8. $0{\cdot}4 \times 100$ ______
9. $\frac{1}{2}$ of $124. ______
10. $\begin{array}{r} 8063 \\ \times \quad 2 \\ \hline \end{array}$

11. 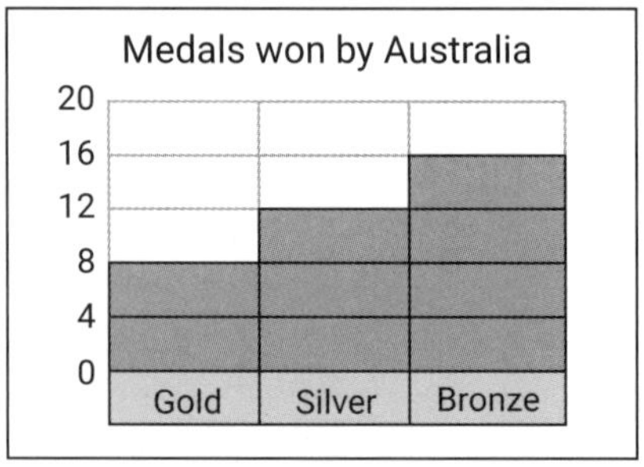 Medals won by Australia

(Bar chart: vertical axis 0, 4, 8, 12, 16, 20; bars Gold, Silver, Bronze)

 a How many of each medal were won?
 G = ______ S = ______ B = ______

 b The type of medal won most. ______

 c How many more Bronze than gold were won? ______

12. In a 200 m race, I fell after running 154 m. How far was I from the finish line? ______

13. A  B 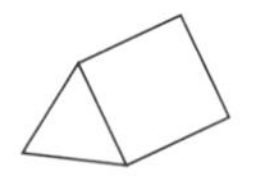 C SOUP D 

 a Which of the objects has a top view that is a circle? ______

 b Which of the objects has a top view that is a square? ______

14. $7\,000\,000 + 50\,000 + 300 + 21 =$ ______

## 3:2 out of 18

1. $7 \times 9 - 38$ ______
2. $8 \times \$62$ ______
3. $21 \div 7$ ______
4. $32 \div 8$ ______
5. $\begin{array}{r} 5735 \\ +\ 2545 \\ \hline \end{array}$
6. Halve 356. ______
7. Multiply $7 by 8. ______
8. $\square \times 7 = 42, \square =$ ______
9. $\square \times 9 = 63, \square =$ ______
10. $\begin{array}{r} 6347 \\ \times \quad 8 \\ \hline \end{array}$

11. Write a digital label for each time.

 a 
 morning
 ___ : ___

 b 
 evening
 ___ : ___

12. $2 \times 3 \times 6 =$ ______
13. $87 - 35 = 92 -$ ______
14. Use am or pm to write the time $1\frac{1}{2}$ hours before:

 a 16:32 ______ b 19:09 ______

 c 09:23 ______ d 05:58 ______

15. Complete this pattern:

| Pentagons | 1 | 2 | 3 | 4 | 5 |
|---|---|---|---|---|---|
| Sides | 5 | | | | |

16. How many $10 notes have the same value as $150? ______
17. Alana is 27 years older than Flynn. How old will Alana be when Flynn is 18? ______
18. $167 + 354 =$ ______ $+ 350$

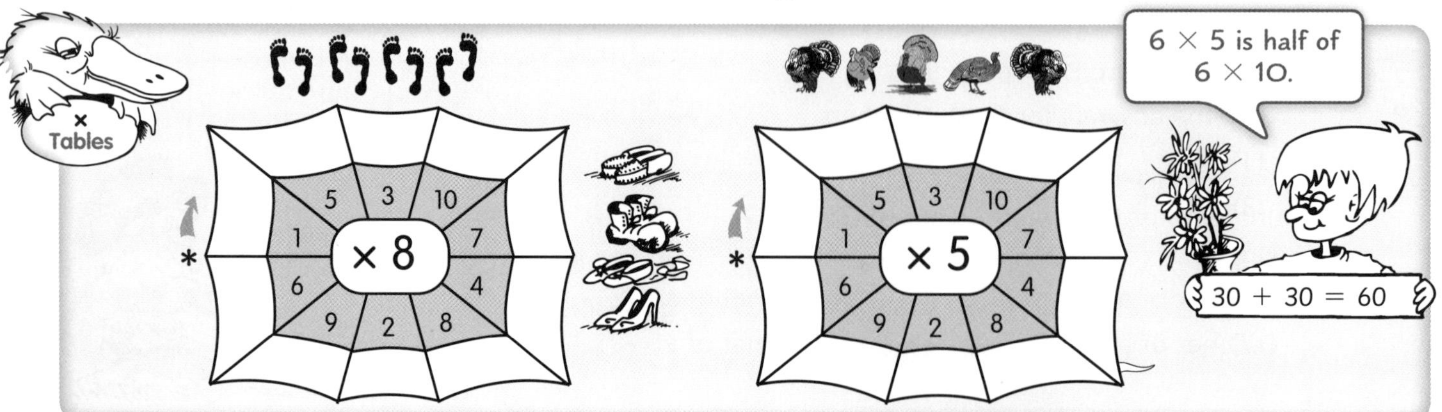

## 3:3 out of 13

1. $5\overline{)830}$
2. $7\overline{)825}$
3. $9\overline{)846}$
4. 28 pants, 30 pairs of shoes. How many more shoes than pants? ______
5. 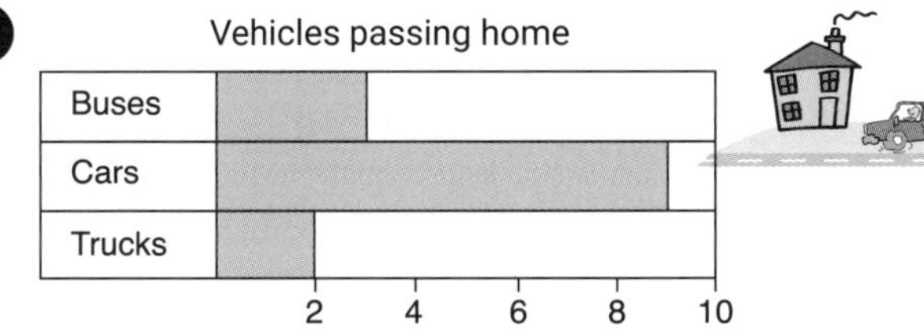

How many vehicles passed my home altogether? ______
6. Rachel is 43 when Lachlan is 5. How old will Rachel be when Lachlan is 37? ______
7. What is the average of 14, 12, 16 and 23? ______
8. Write in order from largest to smallest: 0·67 1·2 0·8 0·99 ______
9. True or false? 700 − 428 = 699 + 1 − 428 ______
10. 8 + 8 + 8 + 8 + 8 + 8 + 8 ______
11. Label each angle with obtuse, acute or reflex.
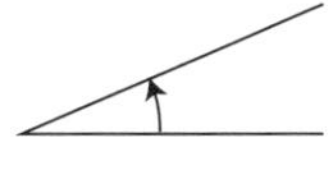 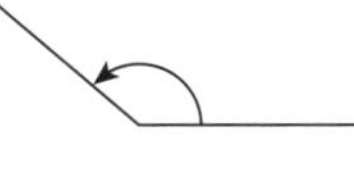 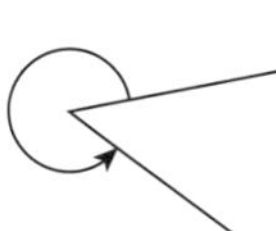
______ ______ ______
12. Write the numeral four thousand and two point two one. ______
13. Make the smallest possible 7-digit whole number using the digits 3, 6, 0, 4, 3, 2 and 9. ______

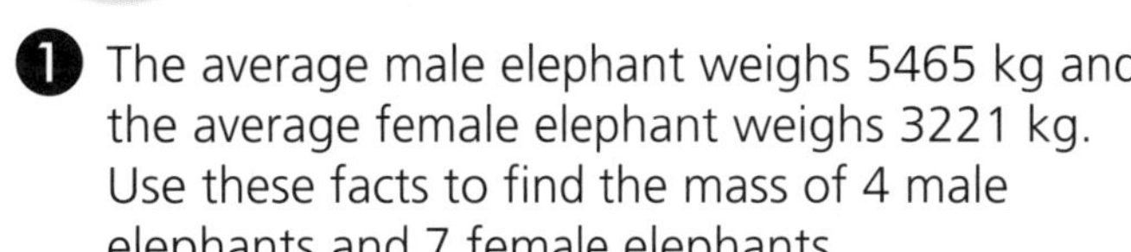


## 3:4 out of 7

**Extension**

1. The average male elephant weighs 5465 kg and the average female elephant weighs 3221 kg. Use these facts to find the mass of 4 male elephants and 7 female elephants. ______
2. What is:
   a one third of half an hour? ______
   b two thirds of half an hour? ______
3. For this solid, calculate the number of corners plus the number of faces minus the number of edges. ______

4. I have 8 boxes. Each box has 4 or 5 books in it. There were 35 books altogether. How many boxes had 4 books? ______
5. I bought at least one of each of these items. The cost was $3.85. How many apples did I buy? ______

6. Use the jump strategy to find 856 + 425.
856
7. 567 + 286 − 186 + 297 ______

**Challenge**

*Write questions that are equal to:*

| a 748 + 268 | b 274 + 377 | c 2978 + 5382 |
|---|---|---|
| = ______ | = ______ | = ______ |
| = ______ | = ______ | = ______ |
| = ______ | = ______ | = ______ |
| = ______ | = ______ | = ______ |
| = ______ | = ______ | = ______ |

To round off to the nearest 5 cents, give the closest answer that ends in 5 or 0.

a Is $37.47 closer to $37.45 or $37.50? ______

Round each of these to the nearest 5 cents.

| | | | | | |
|---|---|---|---|---|---|
| b | $43.88 ______ | c | $29.99 ______ | d | $84.63 ______ |
| e | $117.12 ______ | f | $265.14 ______ | g | $630.97 ______ |
| h | $365.28 ______ | i | $289.09 ______ | j | $836.34 ______ |

## 4:1   ☐ out of 22

1. $8 \times 2$ ____
2. $8 \times 4$ ____
3. $32 \div 8$ ____
4. $16 \div 2$ ____
5. $\begin{array}{r} 2564 \\ +\ 5463 \\ \hline \end{array}$
6. Take 4 from 800. ____
7. Multiply 5 by 6. ____
8. $0{\cdot}5 \times 100$ ____
9. Double 423. ____
10. $\begin{array}{r} 6208 \\ \times\ \ \ \ 3 \\ \hline \end{array}$
11. Write all the factors of 15. ____
12. The first 10 multiples of 6 are: ____
13. **a** 4 squared ____ **b** 8 squared ____
14. These 21 toys are equally shared between 7 children.
    One share = ____

15. **a** $3 \times$ ____ $= 27$ **b** $6 \times$ ____ $= 42$
16. 4 baskets with 20 eggs each. ____ eggs
17. How many days in 5 weeks? ____
18. Haley cut 15 cm from a 1 m ruler. How much of the ruler remained? ____
19. Is a population of 2 706 000 closer to 2 000 000 or 3 000 000? ____
20. Complete the labels for the shaded section.
    ____ tenths or 0· ____

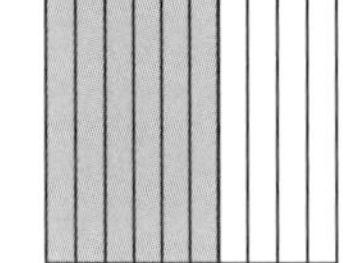

21. Write fifty-one hundredths as a decimal. ____
22. How many thousandths in 0·364? ____

## 4:2   ☐ out of 20

1. 5 squared ____
2. $6 \times 8 - 46$ ____
3. $48 \div 8 + 3$ ____
4. $21 \div 7 \times 9$ ____
5. $\begin{array}{r} 8354 \\ -\ 2746 \\ \hline \end{array}$
6. Years in 4 decades. ____
7. Months in 6 years. ____
8. Minutes in 4 hours. ____
9. Years in 3 centuries. ____
10. $\begin{array}{r} 7463 \\ \times\ \ \ \ 8 \\ \hline \end{array}$
11. 36 cards shared between 4 people.
    One share = ____
12. **a** $6 \times$ ____ $= 42$ **b** $49 \div 7 =$ ____
13. How many 50c apples can be bought for \$3.70? ____
14. I poured 375 mL out of a full 3 L container.
    How much was left in the container? ____
15. 18 shoes are in a shop window.
    How many pairs are there? ____

16. Each hat has 4 corks attached.
    **a** How many corks are on 6 hats? ____
    **b** How many hats could be made with 32 corks? ____

17. How many minutes in 8 hours? ____
18. How many are left over if 22 toys are shared by:
    **a** 3 girls? ____ **b** 4 girls? ____
19. Write 3 out of 10 as a:
    **a** decimal ____ **b** fraction $\frac{\square}{\square}$

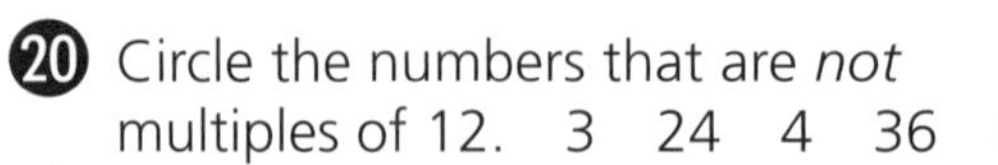

20. Circle the numbers that are *not* multiples of 12. 3 24 4 36 6

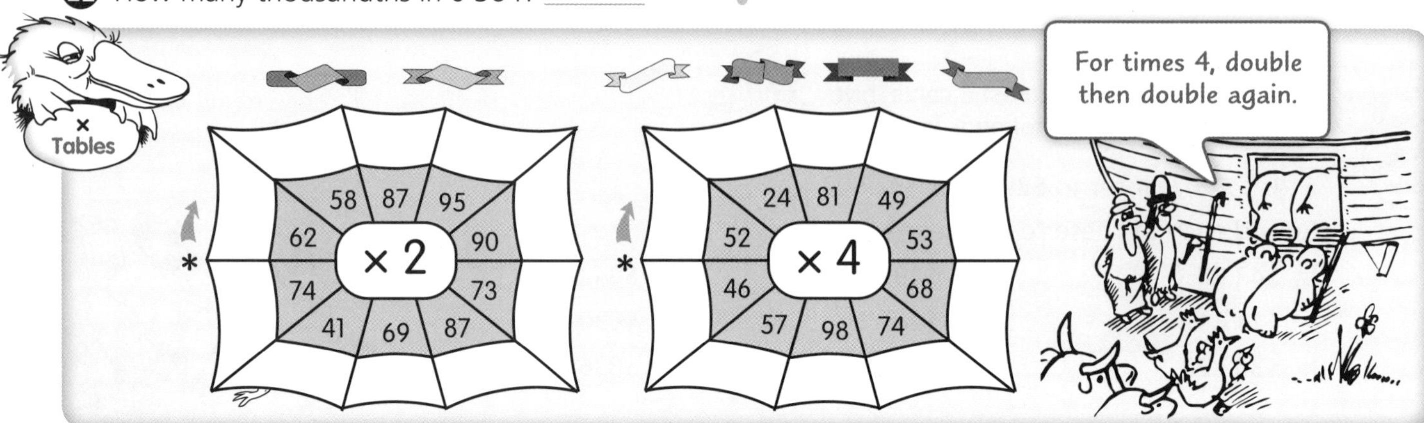




 ISBN 978 0 6557 0886 5

## 4:3 ☐ out of 13

1. $8\overline{)872}$  2. $3\overline{)813}$  3. $4\overline{)724}$
4. Scott was 44 when Felicity was 7.
   How old will Scott be when Felicity is 36? ______
5. Write all the factors of 24. ______
6. The first 10 multiples of 8 are:
   ______
7. **a** 3 squared ______ **b** 9 squared ______
8. I shared 36 bananas in bags of 6.
   How many bags of 6 bananas do I have? ______

9. What is the smallest even number that is a multiple of 9? ______
10. **a** 8 × ____ = 56 **b** 49 ÷ 7 = ____
11. **a** Two emus were 1·75 m and 1·89 m tall. What was the difference between their heights? ______

    **b** The emus had a mass of 42·89 kg and 53·92 kg. What was their total mass? ______
12. How many different straight lines of 3 circles can be found on this picture? ______
    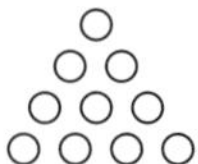
13. Two darts are thrown into this dartboard. Which totals (below 16) are impossible to obtain?
    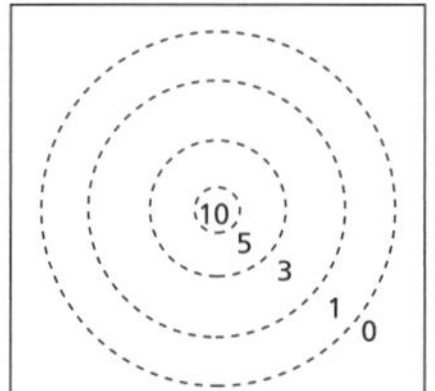
    
    ______

## 4:4 ☐ out of 7

**Extension**

1. After opening a book, the sum of the two page numbers I could see was 145. What were the page numbers? ______
2. Halve the number that is 53 bigger than 177.
3. In one week, Chloe watched 14 h 30 min of TV. Jenna watched 10 h 45 min. How much did they watch altogether? ______

4. 12 boxes of pens held either 5 or 6 pens. There was a total of 64 pens. How many boxes held 6 pens? ______
5. Four of these cockatoos have an average mass of 536 g. The other four have an average mass of 723 g. What is the total mass using these facts. ______

6. **a** What is the sum of the first four square numbers? ______
   **b** What is the product of the first three square number? ______
7. A book has 194 pages. How many times was the digit 8 used in numbering its pages? ______

**Challenge**

*Write questions that are equal to:*

**a** 894 − 283 **b** 674 − 276 **c** 7685 − 4526

| a | b | c |
|---|---|---|
| = ______ | = ______ | = ______ |
| = ______ | = ______ | = ______ |
| = ______ | = ______ | = ______ |
| = ______ | = ______ | = ______ |
| = ______ | = ______ | = ______ |

## 5:1 　 ☐ out of 21

1. 3 × 7 ______
2. 3 × $4 ______
3. 24 ÷ 8 ______
4. 50 ÷ 10 ______
5. $\begin{array}{r} 2967 \\ -\ 1974 \\ \hline \end{array}$
6. 2 squared. ______
7. 24 divided by 6. ______
8. Multiply 7 by 5. ______
9. 0·7 × 100 ______
10. $\begin{array}{r} 2648 \\ \times\ \ \ \ 3 \\ \hline \end{array}$
11. Write all the factors of 12. ______
12. How many groups of 4 in 12 apples.
13. Equal numbers of birds are put in four cages. If there are 24 birds, how many are in each cage? ______

14. **a** 5 × ____ = 60 　 **b** 32 ÷ 8 = ____
15. Write in ascending order. 31 673 316 37 631 713 33 761 301 31 763 116

______

______

16. **a** The difference between 83 and 69. ______

    **b** The total of 465 and 39. ______

    **c** 34 less than 100. ______
17. Thomas drew 4 monsters. He gave each one 7 legs. How many legs altogether? ______
18. The numeral for $10^2$? ______
19. Count on from 76 to find 85 − 76. ______
20. What is 0·98 as a percentage? ______
21. $(7 \times 10^3) + (2 \times 10^2) + (7 \times 10^1) + 7$ ______

## 5:2 　 ☐ out of 18

1. 6 × 9 ______
2. $\frac{4}{6} - \frac{1}{3}$ ______
3. $40 ÷ 8 ______
4. 36 ÷ 4 ______
5. $\begin{array}{r} 3967 \\ +\ 4786 \\ \hline \end{array}$
6. 5 squared ______
7. Multiply 9 by 5. ______
8. Divide $36 by 4. ______
9. 0·9 × 1000 ______
10. $\begin{array}{r} 2945 \\ \times\ \ \ \ 6 \\ \hline \end{array}$
11. The first 10 multiples of 7 are:

______

12. 

    For these trees, how many groups of:

    **a** 6 trees? ______ groups ______ left

    **b** 5 trees? ______ groups ______ left
13. **a** 9 × ____ = 63 　 **b** 81 ÷ 9 =
14. What are the next two odd numbers after 30?

______
15. Round off each number to the nearest hundred, then use these to estimate:

    **a** 873 − 204 ______ 　 **b** 1907 − 743 ______
16. A man's step is 60 cm long. How far does he walk in 7 steps? ______
17. Count on from 157 to find 165 − 157. ______
18. Draw lines to join equivalent numbers.

| | |
|---|---|
| 0·61 | 80% |
| 0·49 | 61% |
| 0·80 | 25% |
| 0·10 | 49% |
| 0·25 | 10% |

**a** Colour 34% red.
Colour 27% blue.
Colour 19% yellow.
What percentage is left uncoloured?

______

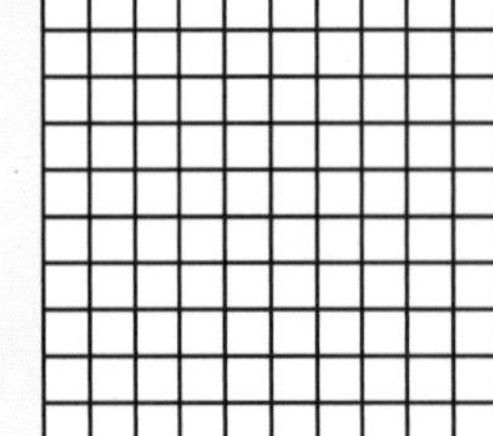

**b** Colour 9% orange.
Colour 41% green.
Colour 38% purple.
What percentage is left uncoloured?

______

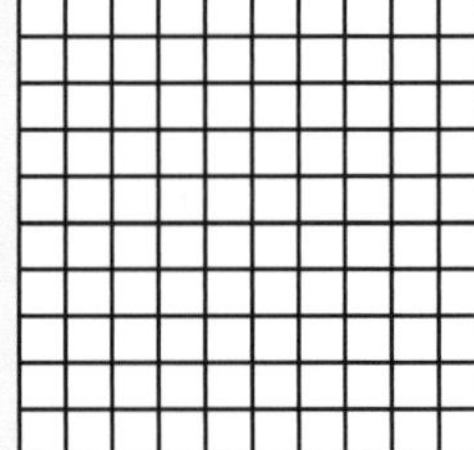

## 5:3 ☐ out of 15

1. $5\overline{)720}$
2. $3\overline{)927}$
3. $7\overline{)719}$
4. $7465 + 1598$
5. $4657 \times 7$
6. 27 exercise books shared between 5 students.

   Books each = ______

   Remainder = ______

7. Write all the factors of 36.

   ______

8. a $6 \times$ ___ $= 42$ b $49 \div 7 =$ ___
9. Which of 1, 15, 23, 49, 60 are:

   a multiples of 5? ______

   b square numbers? ______
10. If one cake will serve 7 people, how many are needed to serve 100? ______
11. Carlos has a part-time job. He is paid $148 per week. From this amount $16 is deducted as tax. How much does he receive for 3 weeks work?
12. Elizabeth has 526 stamps. Max has 8 times as many.

   a How many stamps does Max have? ______

   b How many do they have altogether? ______
13. Write a number sentence for 114 more than a certain number gives 539.

   ______
14. What percentage of a metre is 1 cm? ______
15. The value of 8 in 68 495 921. ______

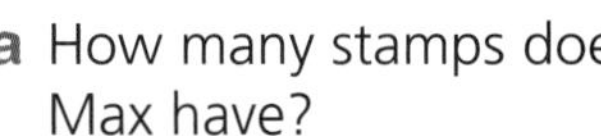

## 5:4 Extension ☐ out of 8

1. What is the smallest square number that is also a multiple of 8. ______
2. Which is larger: $2^2 + 3^2 + 4^2$ or $1^2 + 5^2$? ______
3. Copy the drawing on the left.

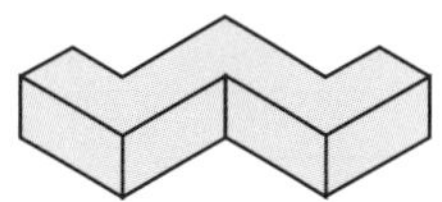

4. ☐ $\div 2 = 55$ ☐ = ___
5. The shaded part has a value of 30. What is the value of the whole? ______

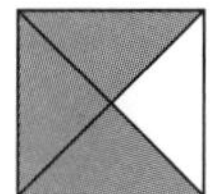

6. If 4 small squares make a quado and 2 quados make an octo, could 68 small squares make:

   a 7 quados and 5 octos? ______

   b 4 quados and 7 octos? ______
7. I am paid between $12 to $14 an hour. Which could be my pay for 6 hours of work: $70.50, $71.80, $83.10 or $84.10? ______
8. How many days in 42 weeks? ______

### Challenge

*List as many square numbers as you can using number sentences, e.g. $2 \times 2 = 4$, 2 squared = 4 or $2^2 = 4$.*

______

______

______

______

Complete this table for the shaded part.

| | | | | |
|---|---|---|---|---|
| [grid] | $\frac{}{10}$ | $\frac{}{100}$ | 0·__ | __% |
| [grid] | $\frac{}{10}$ | $\frac{}{100}$ | 0·__ | __% |

## 6:1 ☐ out of 19

1. 5 × 7 + 12 ____
2. 6 × 2 − 8 ____
3. 20 ÷ 4 + 1 ____
4. 36 ÷ (6 + 3) ____
5. 3574 + 2768
6. 36 − (6 + 3) ____
7. 41 − 16 + 8 ____
8. 46 + ____ = 100
9. 35 + ____ = 82
10. 1423 × 3

11. 81% means ____ out of ____.
12. Complete the labels for the shaded part.

| . | $\frac{}{100}$ | % |
|---|---|---|

13. Arrange in descending order:
67 541 234, 67 647 324, 67 747 324
____
14. Use numerals to write ninety-five million seven hundred and forty-seven thousand and thirteen.
____
15. Is 35 465 798 closer to 35 000 000 or 36 000 000? ____
16. **a** 3 × 4 × 2 ____ **b** 60 − 40 + 23 ____
17. What 2D shape has 4 right angles and has its opposite sides equal? ____
18. Name these 3D shapes.

**a**  **b** 

19. How many faces on three cubes? ____

## 6:2 ☐ out of 19

1. 9 × 8 − 35 ____
2. 6 squared ____
3. 7 × ____ = 56
4. 4 × ____ = 24
5. 4675 − 2798
6. $\frac{9}{12} - \frac{2}{6}$ ____
7. Increase 465 by 67. ____
8. 167 + 45 = ____ + 50
9. 534 + ____ = 546 + 160
10. 3647 × 4
11. Write the value of the 6 in 35 674 725. ____
12. 76 out of 100 = $\frac{76}{100}$ = ____.____ = ____ %
13. $(5 \times 10^4) + (3 \times 10^3) + (8 \times 10^2) + 3$ ____
14. Is this a reflection, translation or rotation? ____

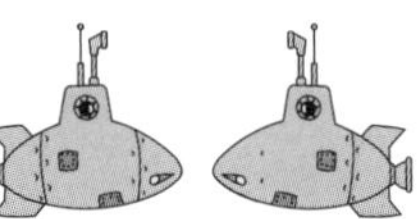

15. Write all the factors of 42.
____
16. Is 72 576 098 closer to 72 000 000 or 73 000 000? ____
17. **a** 8 + (4 × 3) − 8 ____ **b** 5 × 4 + 9 ____
18. Which of these nets could not fold to make a solid shape like this?

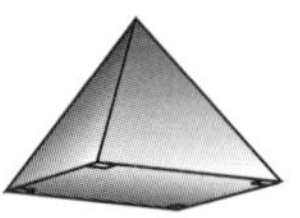

**A** 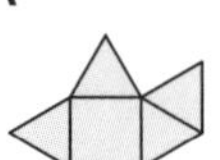 **B**  **C** 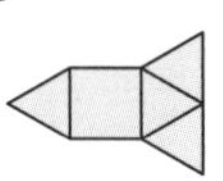 ____

19. Reflection, translation or rotation?

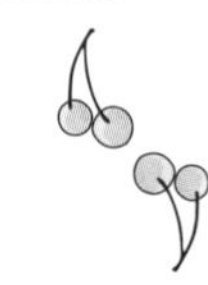

____

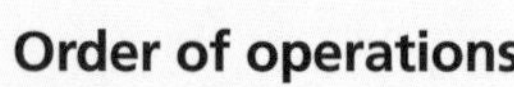

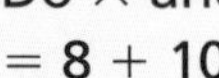

Concept

**Order of operations**

**❶ ( ) ❷ × and ÷ ❸ + and −**

**Example**

4 × (11 − 9) + 20 ÷ 2
Remove the ( ).
= 4 × 2 + 20 ÷ 2
Do × and ÷, (left to right).
= 8 + 10
= 18

**a** 11 − (8 − 3) ____ **b** 14 − (20 − 10) ____
**c** 8 + 2 × 4 ____ **d** 16 − 2 × 6 ____
**e** 20 − 12 ÷ 4 ____ **f** 15 + 6 ÷ 3 ____
**g** 6 ÷ 3 × 2 ____ **h** 20 ÷ 5 × 4 ____
**i** 11 − 4 + 5 ____ **j** 21 − 11 + 6 ____
**k** 63 + 12 ÷ 6 − (8 + 12) ÷ (9 − 4 + 5) ____

## 6:3 out of 14

1. $8\overline{)368}$
2. $6\overline{)936}$
3. $4\overline{)275}$
4. $\begin{array}{r} 6846 \\ -\ 4709 \\ \hline \end{array}$
5. $\begin{array}{r} 3809 \\ \times\ \ \ 5 \\ \hline \end{array}$
6. If one pizza is needed to serve 3 people, how many pizzas are needed to serve 23 people? ______

7. Draw lines to connect equal numbers.

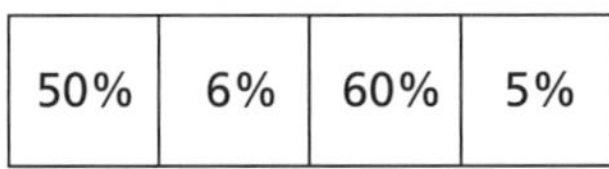

| 0·6 | $\frac{1}{2}$ | 0·06 | $\frac{5}{100}$ |
|---|---|---|---|

| 50% | 6% | 60% | 5% |
|---|---|---|---|

8. Write as a percentage:
   **a** 0·35 ______ **b** $\frac{1}{4}$ ______
9. $(5 \times 10^4) + (6 \times 10^3) + (4 \times 10^2) + (9 \times 10^1) + 9$
   = ______
10. Is 89 465 736 closer to 89 000 000 or 90 000 000? ______
11. Write the value of 2 in 45 627 536. ______
12. **a** $24 \div 6 \times 5 \div 4$ ______ **b** $6 \times 5 - 17$ ______
13. Draw a triangular pyramid. How many:
    **a** faces? ______
    **b** edges? ______
    **c** vertices? ______

14. What shape can be made from this net?

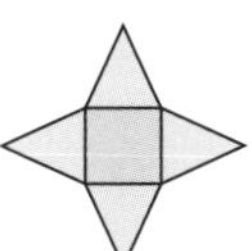

______

## 6:4 Extension out of 6

1. The shortest distance by road from:
   **a** **A** to **F** ______ **b** **C** to **D** ______

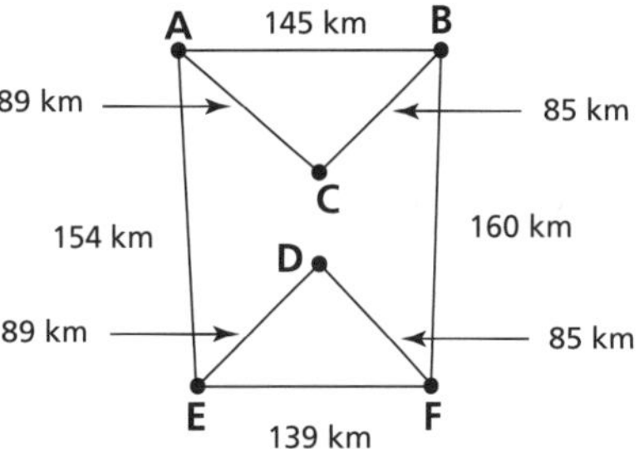

2. If 3 small squares make a *trio* and 3 *trios* make a *nino*, could 57 small squares make:

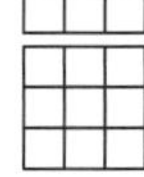

   **a** one *trio* and 6 *ninos*? ______
   **b** 7 *trios* and 4 *ninos*? ______
3. Which number below is a multiple of 5, 6 and 7?
   1764 1470 1120 1560 ______
4. $756 + 546 + 293 - 93$ ______
5. $(5 \times 37) + (5 \times 37)$ ______
6. What two numbers have a difference of 3 and a product of 208? ______

### Challenge

Write number sentences that are equal to 14, using 2 or more operations (+, −, × and ÷) and brackets, e.g. $4 + 5 \times 2 - (4 \times 0) = 14$.

______
______
______
______

### Concept

| 4, 45, 16, 21, 23, 100, 35, 36, 81, 49, 14 |
|---|

1. Which of the numbers above are square numbers? ______
2. Which are multiples of:
   **a** 2? ______
   **b** 5? ______
   **c** 7? ______

## 7:1 ☐ out of 16

1. $3 \times 6 + 9$ ______
2. $4 \times 3 + 16$ ______
3. $5 + 2 \times 10$ ______
4. $2 \times 4 + 34$ ______
5. $\begin{array}{r} 7463 \\ +\ 2589 \\ \hline \end{array}$
6. $24 \div 3 + 20$ ______
7. $15 \div (20 - 15)$ ______
8. $60 \div 6 \times 5$ ______
9. $80 \div 10 - 7$ ______
10. $\begin{array}{r} 5163 \\ \times\ \ \ \ 4 \\ \hline \end{array}$
11. On a cube, what is the number of:

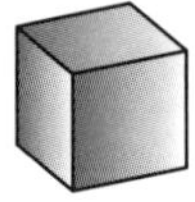

   a faces? ______

   b corners? ______
12. a 4 squared = ______ b $3^2$ = ______
13. Write $1\frac{1}{2}$ as an improper fraction. $\frac{\square}{\square}$
14. Write as a percentage:

   a 0·72 ______ b 0·09 ______
15. Complete this picture so that it is symmetrical about the dotted line.

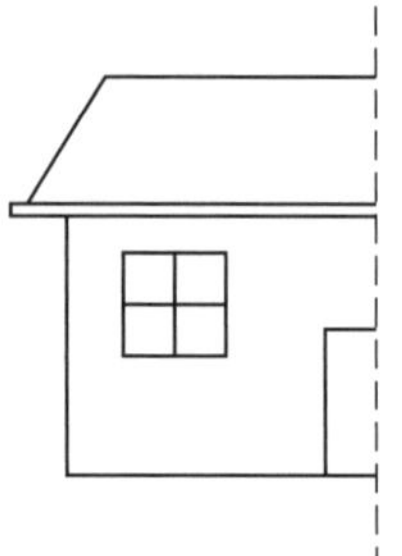

16. Join:
   1 and 5, 5 and 9, 9 and 1,
   2 and 6, 6 and 10, 10 and 2,
   3 and 7, 7 and 11, 11 and 3,
   4 and 8, 8 and 12, 12 and 4.
   What have you drawn?

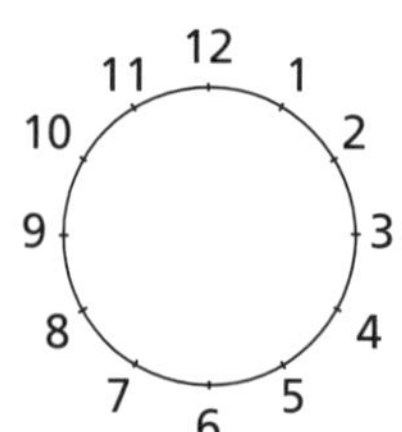


______

## 7:2 ☐ out of 17

1. $9 \times 8 - 65$ ______
2. $57 + 6 \times 7$ ______
3. $45 \div (18 - 9)$ ______
4. $0{\cdot}4 \times 100$ ______
5. $\begin{array}{r} 9054 \\ -\ 3646 \\ \hline \end{array}$
6. $4 \times 7 + 134$ ______
7. $100 \div 10 + 465$ ______
8. $\$37 + 6 \times \$7$ ______
9. $6 \times 4 + (35 \div 7)$ ______
10. $\begin{array}{r} 2074 \\ \times\ \ \ \ 4 \\ \hline \end{array}$
11. Write the percentage equivalent for:

   a $\frac{60}{100}$ ______ b 0·3 ______
12. $(9 \times 10^4) + (6 \times 10^3) + (4 \times 10^2) + 8$ ______
13. For a sphere, how many:

   a surfaces? ______ b edges? ______
14. Name the solid for each net below.

   a

   b

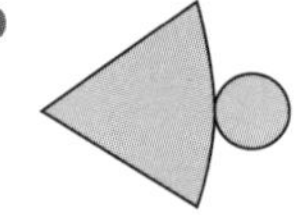

   ______ ______
15. a 8 squared = ______ b $6^2$ = ______
16. Write an improper fraction and mixed number for the shaded part. $\frac{\square}{\square}$ $\square\frac{\square}{\square}$

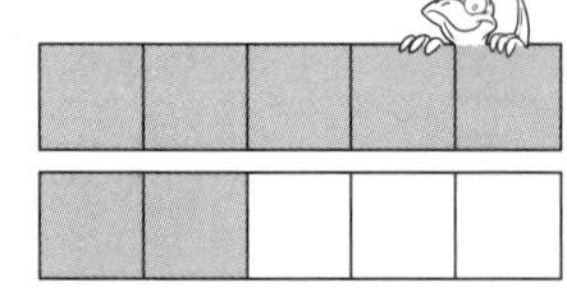

17. Draw a:

   a cube b triangular prism

a How many stakes are needed for a fence 30 m long if the stakes are 5 m apart? ______

b The sum of two numbers is 11 and their product is 24. What are the numbers? ______

c The average of two numbers is 9. If one number is 4, what is the other? ______

 ISBN 978 0 6557 0886 5

## 7:3 out of 12

1. $3\overline{)936}$
2. $6\overline{)408}$
3. $9\overline{)585}$
4. True or false? $0{\cdot}25 = 25\% = \frac{1}{4}$ ______
5. $(9 \times 10^4) + (2 \times 10^3) + (4 \times 10^2) + (8 \times 10^1) + 5$ = ______
6. **a** $9^2 =$ ______ **b** 7 squared = ______
7. Draw a copy of each solid.

**a**

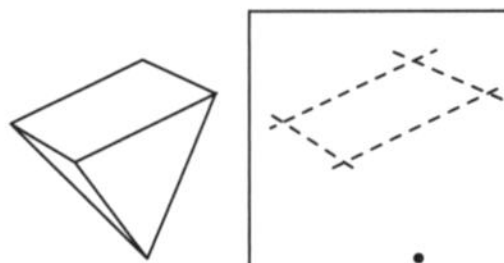

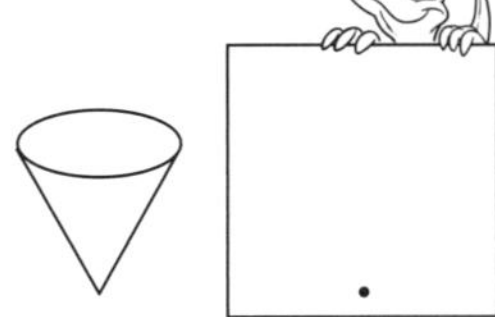

**b**

8. Write $\frac{15}{4}$ as a mixed numeral.
9. Complete:

| Number of years | 1 | 2 | 3 | 4 | 5 |
|---|---|---|---|---|---|
| Number of months | 12 | | | | |

**a** Write a rule to describe the pattern. ______

**b** How many months in 30 years? ______

10. **a** Colour $1\frac{4}{10}$.

**b** Write this as an improper fraction.

11. The improper fraction modelled here.

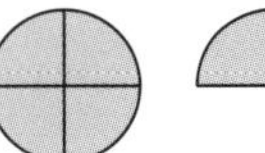

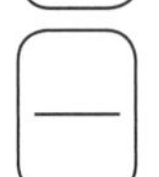

12. Write 0·56 as a:

**a** percentage ______ **b** fraction

## 7:4 Extension out of 5

1. Circle the options that are equal to square numbers.
   49, $2^2 + 3^2$, 16, 9 + 13, 49 − 13
2. **a** Fraction shaded ______
   **b** Decimal shaded ______
   **c** Percentage shaded ______
   **d** Percentage not shaded ______
3. Which of the shapes below can be joined to shape **S** to make a square? ______

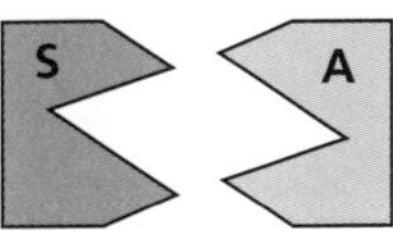



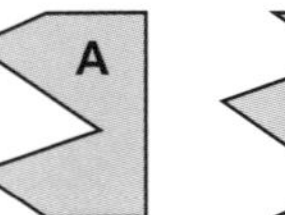



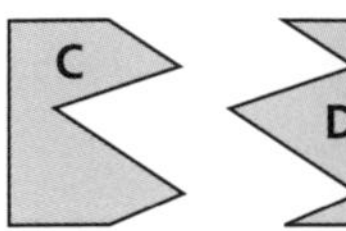



4. 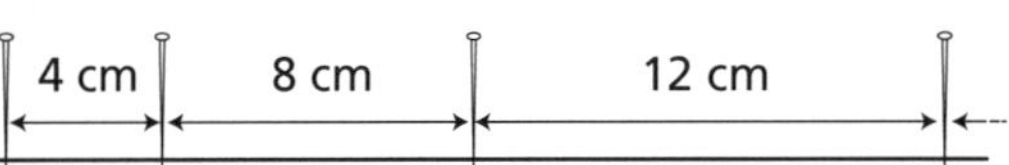



Pins were stuck onto a board using the pattern shown above.
What was the distance between:

**a** the 5th and 6th pins? ______

**b** the 1st and 6th pins? ______

5. If this net is folded to make a cube, which side would be opposite **A**? ______

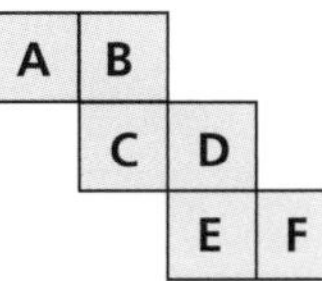



### Challenge

*Write numerals then record them as powers of ten, e.g.*
$8\,724 = (8 \times 10^3) + (7 \times 10^2) + (2 \times 10^1) + 4$

### Reflections

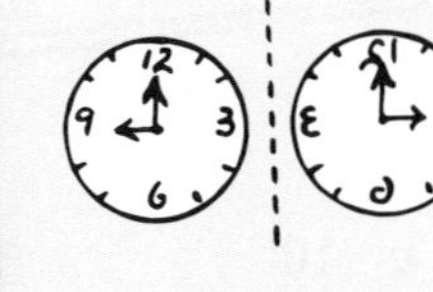

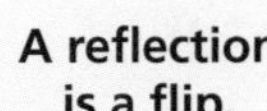

A reflection is a flip.

**a** I saw the clock in the mirror. It looked like 2 o'clock but it was really ______.

Refer to the flip on the left.

**b** Do the lengths of the arms change? ______

**c** Does the angle between arms change? ______

**d** Does the size of the clock change? ______

## 8:1 ☐ out of 17

1. $4 \times 2 + 14$ ______
2. $7 \times 2 + 34$ ______
3. $9 + 6 \times 5$ ______
4. $67 - 5 + 9$ ______
5. $\begin{array}{r} 8603 \\ -\ 2649 \\ \hline \end{array}$
6. $80 \div (45 - 35)$ ______
7. $100 - 9 \times 7$ ______
8. $\$56 + 8 \times \$4$ ______
9. $90 \div (67 - 57)$ ______
10. $\begin{array}{r} 1856 \\ \times \quad 5 \\ \hline \end{array}$
11. Write 0·45 as a percentage. ______
12. **a** $5^2 =$ ______ **b** 6 squared = ______
13. Name this 3D solid. ______

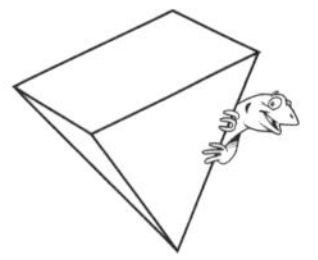

14. **a** $10^2 =$ ______ **b** $7^2 =$ ______
15. Write $\frac{19}{3}$ as a mixed numeral. $\square\frac{\square}{\square}$
16. Which is lower, −5°C or −17°C? ______
17. **Tennis matches won**

| Player | Matches |
|---|---|
| Naomi | ⚾ ⚾ ⚾ ⚾ ⚾ |
| Alana | ⚾ ⚾ ◐ |
| Luke | ⚾ ⚾ ⚾ ⚾ ⚾ |
| Rachel | ⚾ ⚾ ⚾ ◐ |
| Heather | ⚾ ⚾ ⚾ ⚾ ⚾ ⚾ |

⚾ = 4 matches

**a** How many matches were won by Alana? ______

**b** Which player won 14 matches? ______

**c** Which player won more than 20 matches? ______

## 8:2 ☐ out of 20

1. $0{\cdot}8 \times 100$ ______
2. $9 \times (10 - 7)$ ______
3. $20 + 56 \div 8$ ______
4. $\frac{9}{10} + \frac{4}{5}$ ______
5. $\begin{array}{r} 3647 \\ +\ 6809 \\ \hline \end{array}$
6. −10, −9, −8, ______, ______
7. Increase 45 by 89. ______
8. 78 − ______ = 34
9. \$45 + ______ = \$135
10. $\begin{array}{r} 2704 \\ \times \quad 8 \\ \hline \end{array}$
11. $(7 \times 10^4) + (8 \times 10^3) + (3 \times 10^2) + (1 \times 10^1) + 8$ = ______
12. List all the square numbers up to 100. ______
13. Write $\frac{83}{100}$ as:

    **a** a percentage ______ **b** a decimal ______
14. **a** $3^2 =$ ______ **b** 8 squared = ______
15. Write $1\frac{3}{4}$ as an improper fraction. $\frac{\square}{\square}$
16. What is 5°C lower than −14 degrees Celsius? ______
17. Write the factors of 56. ______
18. How many dots would be in the 6th row? ______

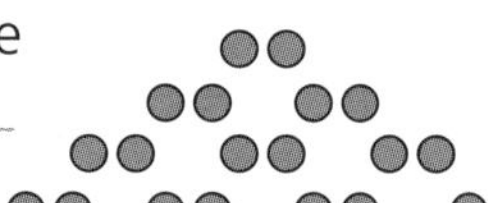

19. Complete this pattern:

| Squares | 1 | 2 | 3 | 4 | 5 |
|---|---|---|---|---|---|
| Sides | 4 | 8 | | | |

20. Complete the missing numbers.

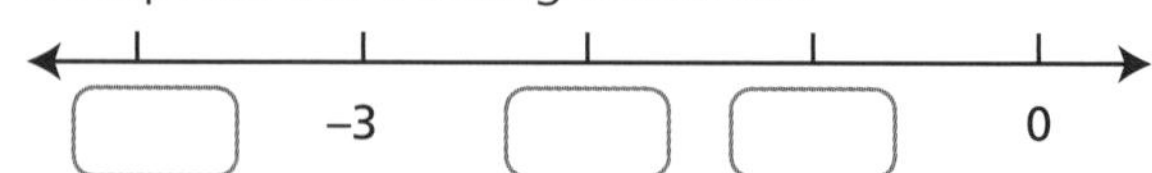

### The Square of a number

To square a number multiply it by itself.

$6 \times 6 = 6$ squared

$6 \times 6 = 6^2$

Find the value of:

**a** 5 squared ______ **b** 3 squared ______ **c** $10^2$ ______

**d** 4 squared ______ **e** $8^2$ ______ **f** $1^2$ ______

**g** 9 squared ______ **h** 2 squared ______ **i** $6^2$ ______

## 8:3 out of 16

1. $4\overline{)596}$
2. $3\overline{)527}$
3. $8\overline{)944}$
4. Write $\frac{19}{5}$ as a mixed numeral.
5. $(2 \times 10^4) + (7 \times 10^3) + (6 \times 10^2) + (7 \times 10^1) + 1$ = ______
6. Complete this pattern.

| 1st number | 50 | 51 | 52 | 53 |
|---|---|---|---|---|
| 2nd number | 65 | 66 | 67 | |

Write a rule for the pattern above. ______

7. Order from smallest to largest: 2, −2, −8, 4, 0 ______
8. 1, 4, 9, 16, ______, ______, ______, ______
9. **a** 0·7 = ______ % **b** 0·1 = ______ %
   **c** 0.07 = ______ % **d** 0·01 = ______ %
10. 807, ______, 819, 825, ______, 837
11. Which of the numbers 9, 14 and 17 have exactly 2 factors? This is called a "prime number". ______
12. Complete the number line.

−0·8 −0·6 −0·4 [ ] [ ]

13. Which is lower, −23°C or −7°C? ______
14. Write the factors of 81. ______
15. Write 38 hundredths as:
    **a** a decimal ______ **b** percentage ______
16. 56 ÷ 7 + (287 − 53) ______

## 8:4 Extension out of 5

1. What is the smallest two-digit square number that is also odd? ______
2. Common means 'belongs to all'. What is the lowest common multiple of:
   **a** 4 and 6? ______ **b** 3 and 5? ______
3. 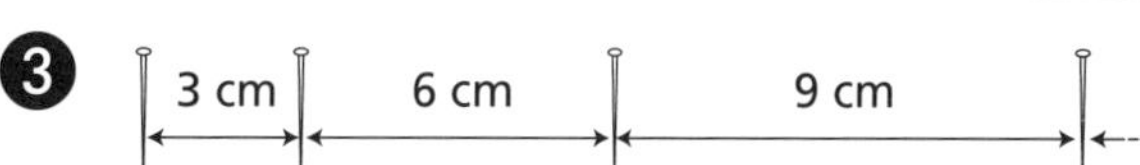

Pins were stuck onto a board using the pattern shown above. What was the distance between:
   **a** the 7th and 8th pins? ______
   **b** the 1st and 8th pins? ______
4. A corner has been cut off this block.
   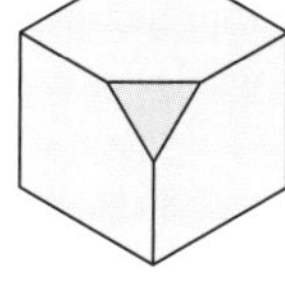
   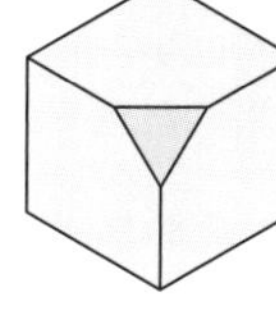
   **a** How many faces would the block have if each of the corners were cut off? ______
   **b** How many edges would it have? ______
5. Complete this table if △ = □ + 39.

| □ | 15 | 25 | 35 | 44 | 54 |
|---|---|---|---|---|---|
| △ | | | | | |

**Challenge**

*Draw and label 3D objects, showing their features.*

**Order of operations**

**Example**

28 − (7 − 3) ÷ 2
Remove the ( ).
= 28 − 4 ÷ 2
Do × and ÷.
= 28 − 2
= 26

Order
**1** ( )
**2** × and ÷
**3** + and −
(going from left to right)

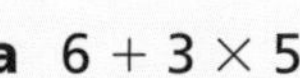

**a** 6 + 3 × 5 ______ **b** 10 − 2 × 4 ______
**c** 7 − (10 − 3) ______ **d** 28 − (20 − 10) ______
**e** 10 − 3 + 4 ______ **f** 10 − (3 + 4) ______
**g** 20 − 4 + 9 ______ **h** 25 − 19 + 1 ______
**i** 20 ÷ 5 × 12 ______ **j** 20 ÷ (5 × 2) ______
**k** 10 + (6 ÷ 2) × 3 + 15 ÷ 3 ______

## 9:1 out of 16

1. $3 \times 8 + 53$ ______
2. Halve 264. ______
3. $100 - 4 \times 10$ ______
4. $15 +$ ______ $= 58$
5. $\begin{array}{r} 5473 \\ -\,2547 \\ \hline \end{array}$
6. $100 \div (2 \times 5)$ ______
7. $89 - 5 \times 4$ ______
8. $72 ÷ 8 ______
9. $9 \times$ ______ $= 45$
10. $\begin{array}{r} 5368 \\ \times \quad 5 \\ \hline \end{array}$
11. Complete the number line.

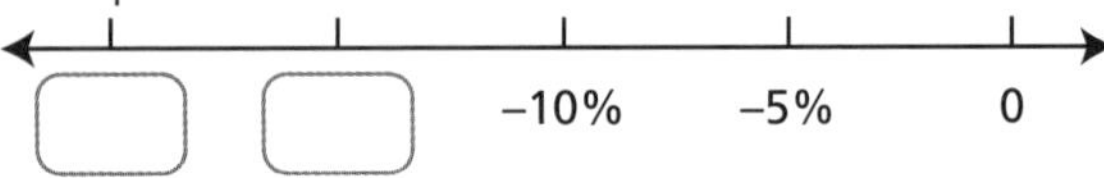


12. Which is lower, $-2$°C or $-12$°C? ______
13. 

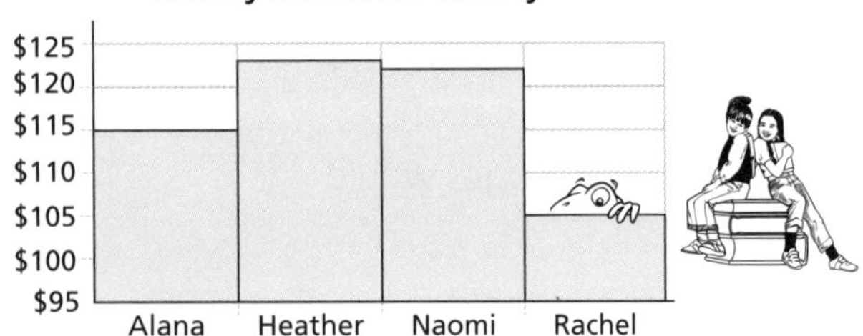


   a How much did Rachel raise? ______
   b How much did these girls raise together? ______
14. Which of these angles is:
   a acute? ______
   b obtuse? ______

15. Which is lower, $-7$ or $-2$? ______
16. C –9 –8 A –6 –5 B –3

   What number would be at:
   a A? ______ b B? ______ c C? ______

## 9:2 out of 18

1. $54 \div 6$ ______
2. $90 ÷ 10 ______
3. $4 \times 19$ ______
4. 9 × $27 ______
5. $\begin{array}{r} \$3576 \\ +\$4797 \\ \hline \end{array}$
6. Increase 76 by 89. ______
7. $56 + 46 =$ ______ $+ 50$
8. $354 - 78 =$ ______ $- 80$
9. 6 less than 4. ______
10. $\begin{array}{r} \$6476 \\ \times \quad 3 \\ \hline \end{array}$
11. $(6 \times 10^4) + (3 \times 10^3) + (9 \times 10^2) + (2 \times 10^1) + 9$ = ______
12. Which is lower, $-15$°C or $-49$°C? ______
13. 

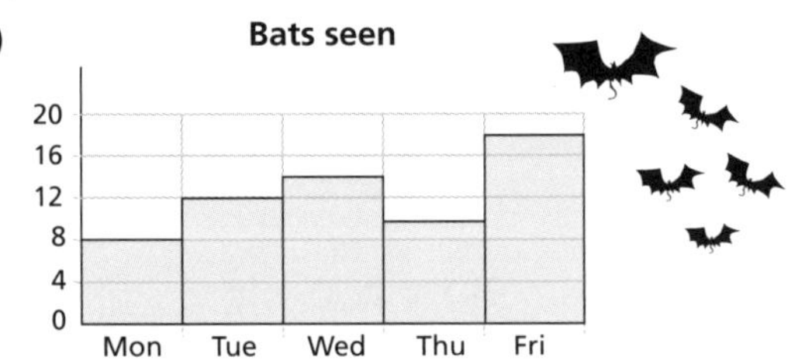


   a How many more bats were seen on Friday than Monday? ______
   b On which two days were the least number of bats seen? ______
14. An angle of size 120° is an ______ angle.

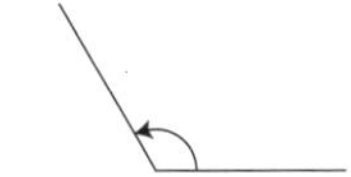

15. Arrange in order of size from lowest to highest.
   $-7, 8, -2, -9, 4$

16. Write $\frac{24}{5}$ as a mixed number. $\square\,\frac{\square}{\square}$
17. $0{\cdot}29 =$ ______ % $= \frac{\square}{\square}$
18. The difference between 87 and 34 is ______.

Turn to ID card B on page 7.
Give the answers for these numbers.

| | | | |
|---|---|---|---|
| (16) | ______ lines | (17) | ______ lines |
| (18) | ______ | (19) | ______ of an angle |
| (20) | ______ angle | (21) | ______ angle |
| (22) | ______ angle | (23) | ______ angle |
| (24) | ______ angle | (25) | ______ |

 • *AUSTRALIAN SIGNPOST MATHS 6 MENTALS* • ISBN 978 0 6557 0886 5

## 9:3 ___ out of 9

1. 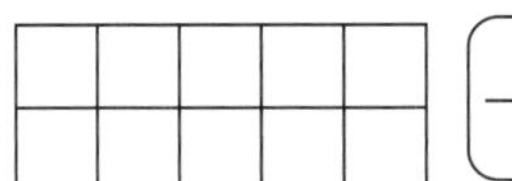 $6\overline{)726}$
2.  $3\overline{)973}$
3. $10\overline{)460}$
4. How many 30° angles make a right angle? ______
5. Write < (less than) or > (greater than) to make these number sentences true.
   - a 4 ______ −5
   - b −8 ______ 3
   - c −3 ______ −21
   - d −7 ______ −1
6. Shade seven tenths and complete the labels.

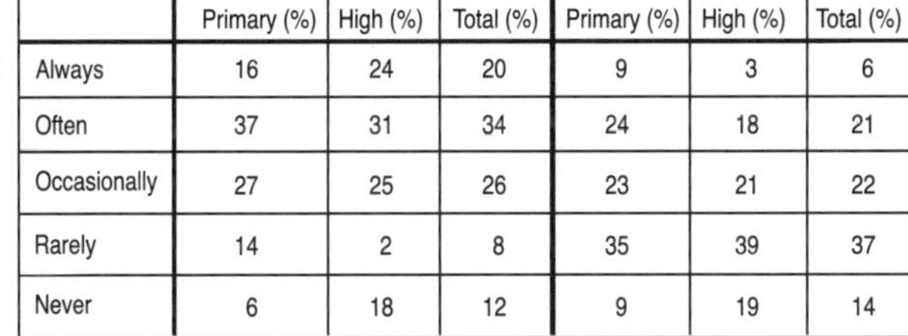

7.

| | Miss breakfast | | | Arrive late | | |
|---|---|---|---|---|---|---|
| | Primary (%) | High (%) | Total (%) | Primary (%) | High (%) | Total (%) |
| Always | 16 | 24 | 20 | 9 | 3 | 6 |
| Often | 37 | 31 | 34 | 24 | 18 | 21 |
| Occasionally | 27 | 25 | 26 | 23 | 21 | 22 |
| Rarely | 14 | 2 | 8 | 35 | 39 | 37 |
| Never | 6 | 18 | 12 | 9 | 19 | 14 |

   - a What fraction of students always miss breakfast? ______
   - b What percentage said they always or often arrive late? ______
   - c If 1000 students were surveyed, how many *never* miss breakfast? ______
8. Write fifty-one hundredths:
   - a as a decimal ______
   - b as a percentage ______
9. 
   - a Colour $\frac{1}{4}$ of this shape red.
   - b Colour $\frac{1}{3}$ of this shape blue.
   - c Is $\frac{1}{3}$ smaller than $\frac{1}{4}$? ______

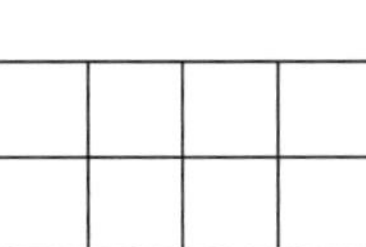

## 9:4 Extension ___ out of 9

1. How many minutes in 2 days and 19 hours? ______
2. How many triangles can be found on this figure? ______

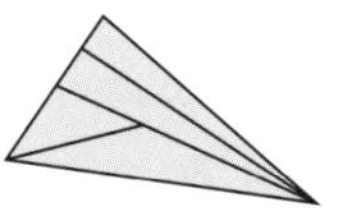

3. Which angle is:
   - a obtuse? ______
   - b acute? ______
   - c straight? ______
   - d reflex? ______

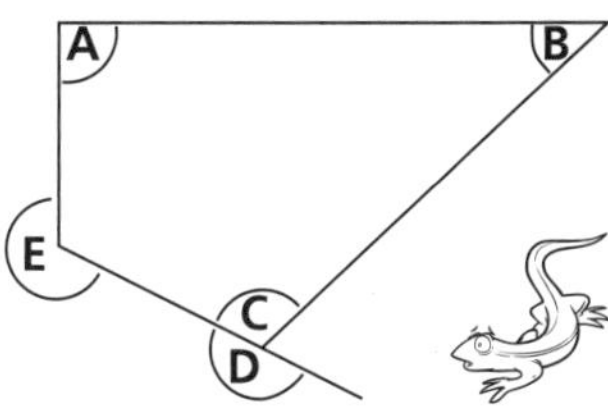


4. $15 \times (87 - 79) - 46 =$ ______
5. What is the angle size between the hands of a clock at 20 past 6? ______
6. How many \$1 coins (2·5 cm wide) are needed side by side to make a length of 27 cm? ______
7. How many octagons of any shape are present? ______

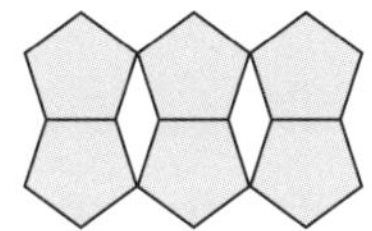

8. Does $2{\cdot}5^2 = 1{\cdot}5^2 + 2^2$?
9. I am able to save \$9 each week. How many weeks will it take me to save \$684? ______

**Challenge**

*Write number sentences that are equal to 56.*

______

______

______

______

______

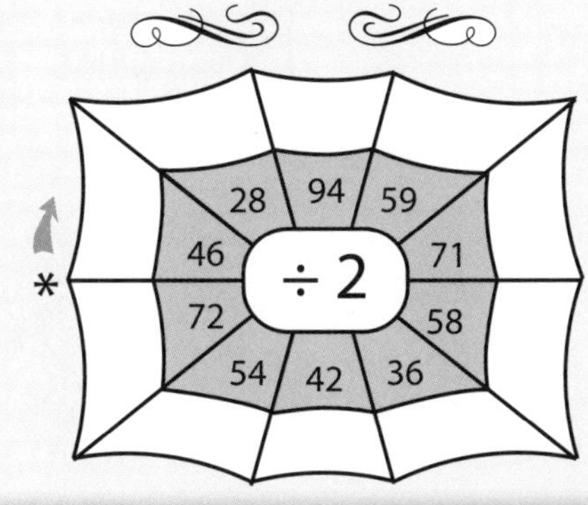

· × 4 = 28, · = ______

## 10:1 out of 16

1. $5 \times (4 + 6)$ ____
2. $78 - 67$ ____
3. $14 - 14 \div 2$ ____
4. $37 - 2 \times 6$ ____
5. $6476 + 1463$
6. $0{\cdot}6 \times 1000$ ____
7. $245 - 35 =$ ____ $- 40$
8. $154 + 34 =$ ____ $+ 30$
9. $7 \times$ ____ $= 56$
10. $6047 \times 4$

11. 
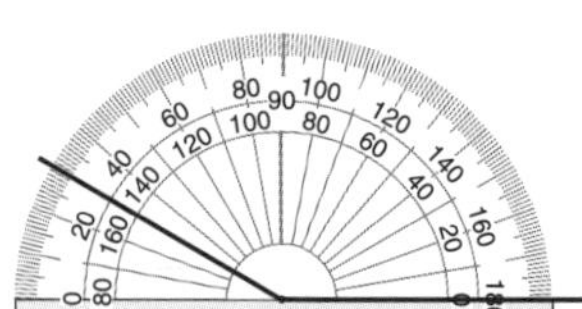

a The size of the angle shown? ____

b Is this angle acute or obtuse? ____

12. Order these numbers from smallest to largest.

56 798 453    56 480 093    56 350 356

____

13. Write 3 ÷ 5 as a fraction, then show the division by shading the diagram.

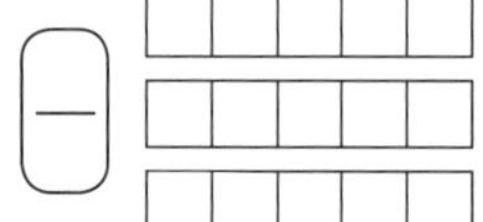

14. a $\frac{1}{5}$ of 15 ____

b $\frac{1}{3}$ of 15 ____

15. C  –4  –3  A  –1  0  B  2

What number would be at:

a A? ____ b B? ____ c C? ____

16. Write the numeral five hundred and three thousand and fifty-four. ____

## 10:2 out of 15

1. $56 + 846$ ____
2. $759 - 79$ ____
3. $2 \times (12 - 6)$ ____
4. $4 \times$ ____ $= 24$
5. $86757 - 36579$
6. $\frac{1}{4}$ of \$12 ____
7. $278 - 57 =$ ____ $- 60$
8. $9 \times (100 - 97)$ ____
9. $64 - 14 - 28$ ____
10. \$6057 × 7

11. How many degrees are in a straight angle? ____

12. C  –80%  –60%  A  –20%  B  20%

What number would be at:

a A? ____ b B? ____ c C? ____

13. Convert to yen:

a 25c ____

b 60c ____

Convert to cents:

c 40 yen ____

d 60 yen ____

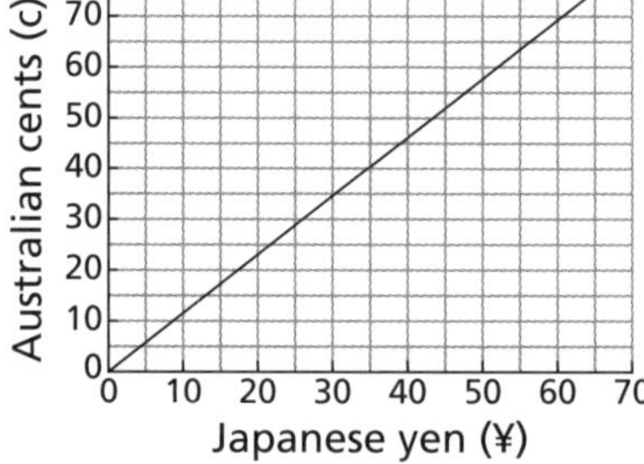


14. a Label what fraction of the hexagon's area is covered by each shape within the hexagon.

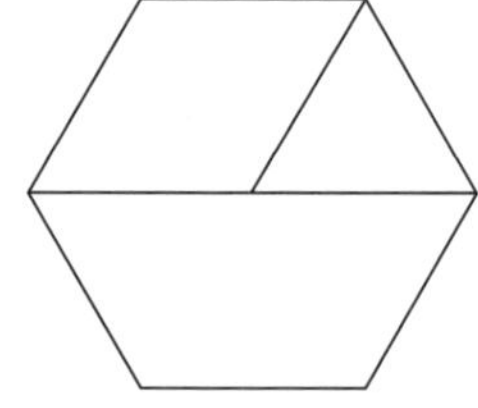

b Is $\frac{1}{3}$ less than $\frac{1}{2}$? ____

c Is $\frac{1}{6}$ more than $\frac{1}{3}$? ____

15. a $\frac{1}{4}$ of 12 ____

b $\frac{1}{3}$ of 12 ____

Turn to ID Card B on page 7.

Give the answers for these numbers.

(1) ____ (2) ____ number

(3) ____ number (4) ____

(5) ____ and ____ (7) ____

(8) ____ numbers (9) ____

(10) ____ ones, ____ tenths, ____ hundredths, ____ thousandths

## 10:3 ☐ out of 10

1. $6\overline{)744}$  2. $7\overline{)847}$  3. $4\overline{)376}$

4. $(3 \times 10^4) + (4 \times 10^3) + (6 \times 10^1) + 7$

   = ______

5. Write $\frac{7}{2}$ as a mixed number. ☐

6. Write < (less than) or > (greater than) to make these number sentences true.

   **a** −3 ______ −7  **b** 6 ______ −5

   **c** 0 ______ − 1  **d** −10 ______ −8

7. Number line: C, −1, $-\frac{1}{2}$, A, $\frac{1}{2}$, B, $1\frac{1}{2}$

   What number would be at:

   **a** **A**? ____  **b** **B**? ____  **c** **C**? ____

8. **a** $\frac{1}{2}$ of 24 ____

   **b** $\frac{5}{6}$ of 24 ____

   **c** $\frac{3}{4}$ of 24 ____

   **d** $\frac{2}{3}$ of 24 ____

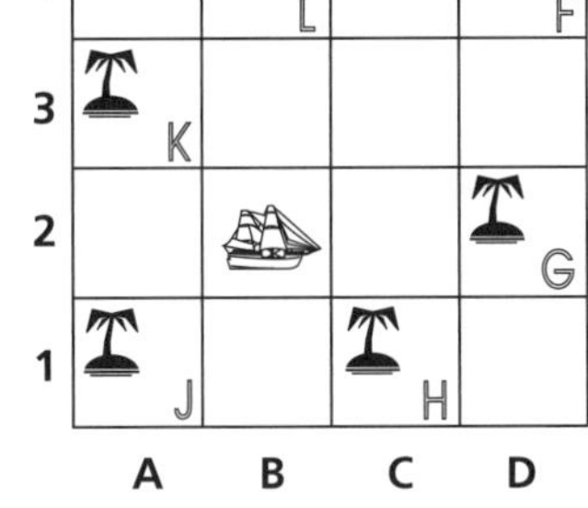

9. Which island is at:

   **a** **D2**? ______

   **b** **B4**? ______

   Which island is:

   **c** north of the ship? ______

   **d** south-east of the ship? ______

|   | A | B | C | D |
|---|---|---|---|---|
| 4 |  | L |  | F |
| 3 | K |  |  |  |
| 2 |  | ship |  | G |
| 1 | J |  | H |  |

10. How many hundreds could be taken from 6 786 453? ______

## 10:4 Extension ☐ out of 5

1. Circle the car that is correctly parked.

______

2. In how many ways can we draw 3 crosses in a row, on this grid? ______

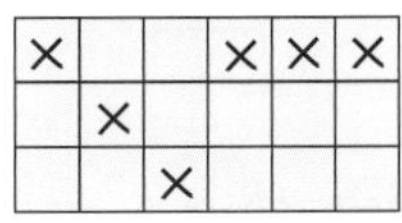

3. Use your ruler and protractor to find:

   **a** the number of equal sides ____

   **b** the number of equal angles ____

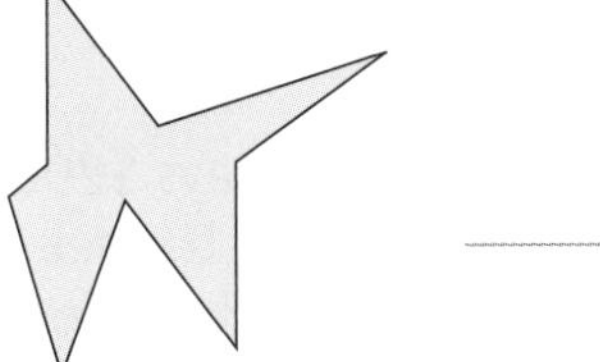

4. **a** $(13 \times 16) + (13 \times 4)$ ____

   **b** $(245 - 15 \times 15) \div 2$ ____

5. One sixth of a whole is 3. Write the value of each shaded part.

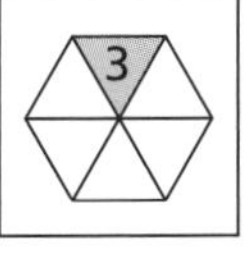


   **a** 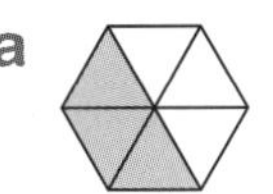 ____

   **b** 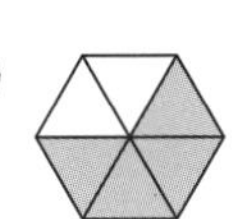 ____

   **c** Find the value of $8\frac{1}{3}$ hexagons. ____

**Challenge**

*Describe this angle. List places you might see an angle this size.*

______

______

______

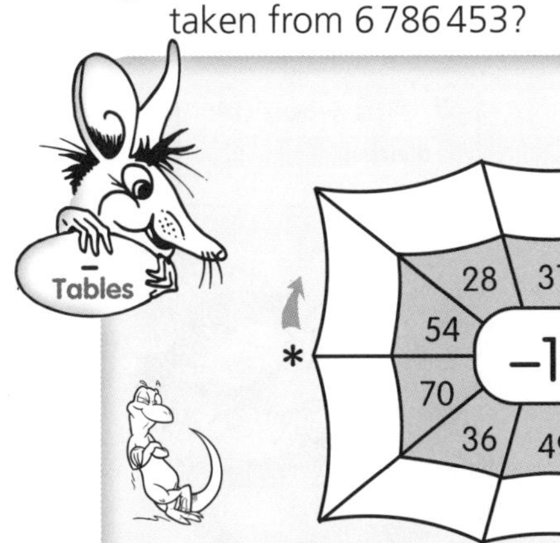


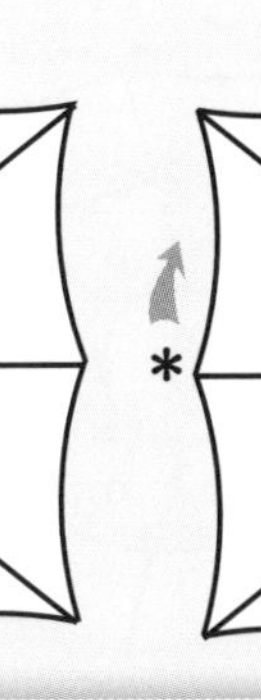

## 11:1 out of 16

1. 60 × 4
2. 3 × 90
3. 156 − 24
4. 0·6 × 10
5. $\begin{array}{r} 8568 \\ -\ 3647 \\ \hline \end{array}$
6. 800 × 3
7. 3 × 500
8. 8 × (56 − 47)
9. $\frac{4}{9} + \frac{3}{9}$
10. $\begin{array}{r} 1404 \\ \times\quad 3 \\ \hline \end{array}$
11. Write 6 ÷ 10 as a fraction, then write the division as a decimal.
12. Our family pays $200 for groceries each week. How much do we pay over 9 weeks?
13. a $\frac{1}{3}$ of 18
    b $\frac{2}{3}$ of 18
    c $\frac{1}{6}$ of 18
    d $\frac{5}{6}$ of 18
14. Which object is:
    a north of **A** and west of **B**?
    b east of **D** and south of **C**?

| | ☆ | C | |
|---|---|---|---|
| D | [flower] | [tree] | B |
| [bird] | | [turtle] | |
| ⑤ | A | | E |

15. $10^3 + 10^2 + 10 + 1$
16. Order these numbers from smallest to largest.
    48 576 386, 48 376 098, 48 265 980

## 11:2 out of 13

1. 2 × 700
2. 700 ÷ 10
3. 0·9 × 100
4. 978 − 59
5. $\begin{array}{r} 3645 \\ +\ 3698 \\ \hline \end{array}$
6. 8 × (97 − 89)
7. $\frac{8}{10} - \frac{3}{5}$
8. 18 + 7 × 10
9. Increase 97 by 18.
10. $\begin{array}{r} 3869 \\ \times\quad 8 \\ \hline \end{array}$
11. a $\frac{1}{3}$ of 18
    b $\frac{1}{6}$ of 18
    c $\frac{5}{6}$ of 18
    d Is $\frac{1}{6}$ less than $\frac{1}{3}$?
12. I bought 6 packets of pasta for $2.95 each. What is the total cost?
13. 

What is found at the coordinates:
a **F3**?
b **J4**?
c **A5**?
d **F5**?
e **C2**?

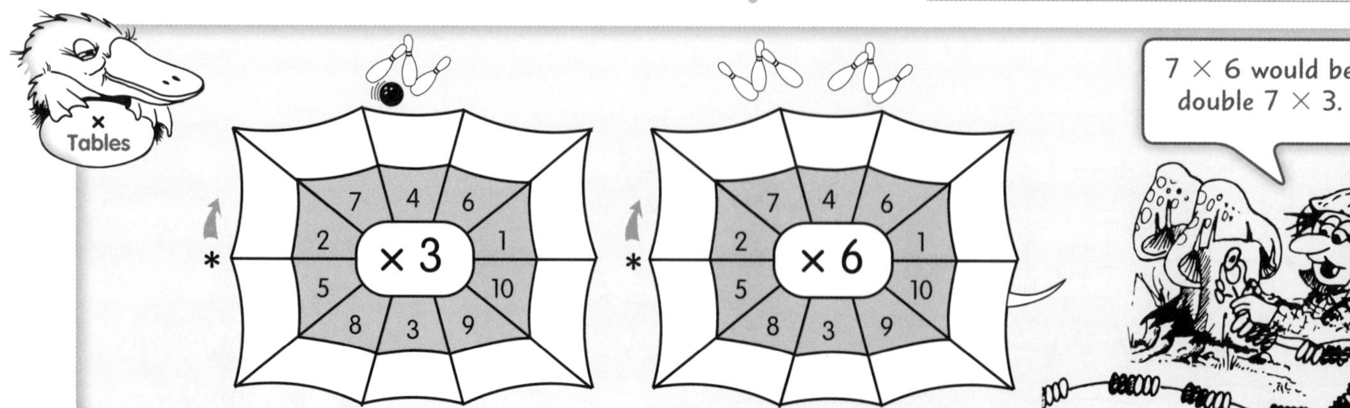



 ISBN 978 0 6557 0886 5

## 11:3 out of 10

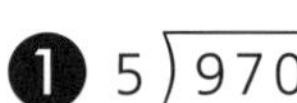

**1** 5)970 **2** 6)468 **3** 7)238

**4** Each day this week, I read an average of 236 pages of my book. How many pages did I read this week? ______

**5** What is the distance from Hobart to:

a **A**? ______

b **B**? ______

c **C**? ______

d **D**? ______

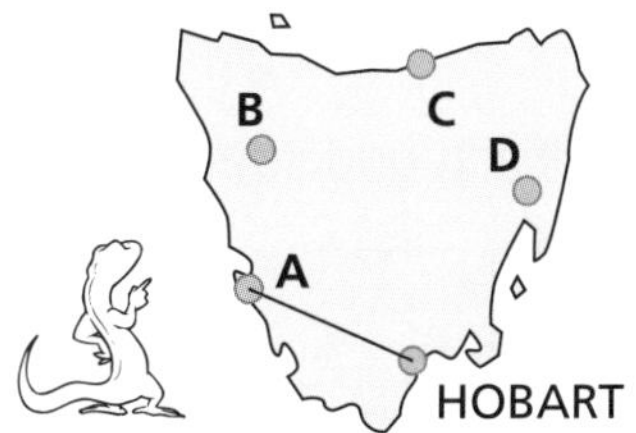


Scale: 1 mm = 18 km

**6** 4 × 7 + 2 + 57 ______

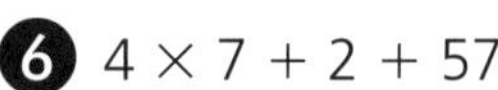

**7** a $\frac{1}{4}$ of 16 ____ b $\frac{3}{4}$ of 16 ____

c $\frac{1}{8}$ of 16 ____ d $\frac{7}{8}$ of 16 ____

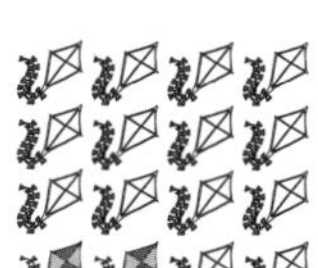

**8**

Give the name of the figure at:

a **E2** ______ b **B1** ______

c **C1** ______ d **D1** ______

e **E1** ______ f **A1** ______

g **D2** ______ h **B2** ______

**9** How many teams of six can be made using 59 children? ______

**10** (80 × 5) + (80 × 5) = ______

## 11:4 out of 6

**Extension**

**1** Two tins of beans are sold for $5.50. Find:

a the cost of 1 tin ______

b the cost of 26 tins ______

**2** How long would it take me to pay for a stove that costs $490 if I pay $99 each week? ______

**3** I can run around the large square once or around the small square four times. What is the difference in distance? ______

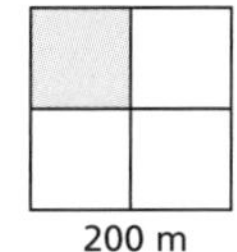


**4** a $6\frac{1}{2} + 13\frac{1}{2}$ ______

b $20 - 3\frac{1}{2}$ ______

**5** Circle the square below that completes the pattern.

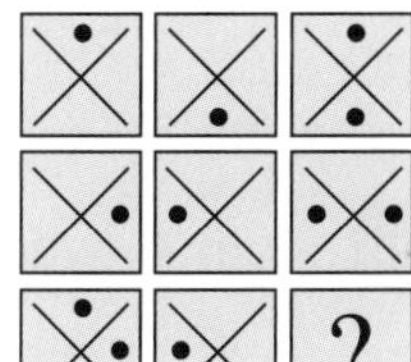


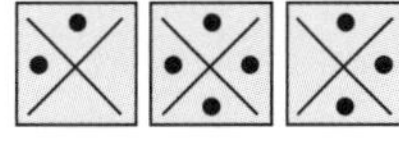

**6** One pizza is enough for 4 people. How many do I need to satisfy 33 people? ______

**Challenge**

*Describe the position of items on this grid, e.g. the pentagon is north-east of the triangle.*

______

______

______

| A | | |
|---|---|---|
| | | B |
| C | | |
| | | |

### Multiplying numbers ending in zeros

× Tables

Write down the end zeros, and multiply the other numbers.

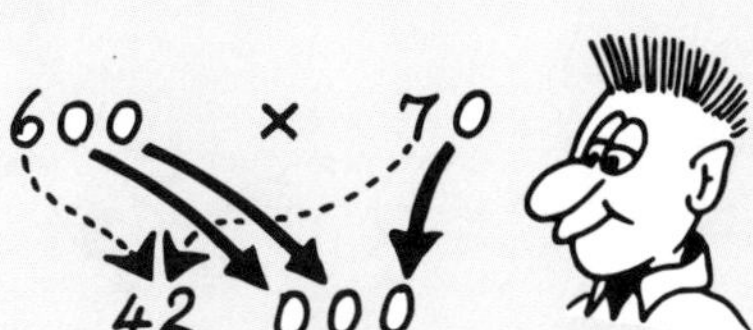


a 50 × 30 ______ b 70 × 80 ______

c 40 × 60 ______ d 50 × 50 ______

e 90 × 50 ______ f 100 × 50 ______

g 80 × 50 ______ h 300 × 40 ______

i $50^2$ ______ j $800^2$ ______

 *AUSTRALIAN SIGNPOST MATHS 6 MENTALS* • ISBN 978 0 6557 0886 5

## 12:1 out of 16

1. 36 + 57 ____
2. $\frac{1}{2} + \frac{1}{2}$ ____
3. 93 − 45 ____
4. 103 − 47 ____
5. $\begin{array}{r} 4657 \\ +\,3987 \\ \hline \end{array}$
6. Multiply 3 by 20. ____
7. Divide 90 by 9. ____
8. 45 + 9 × 4 ____
9. 56 + 87 = ____ + 90
10. $\begin{array}{r} 1024 \\ \times \quad 5 \\ \hline \end{array}$
11. Con carries 6 loads of milk, each with a capacity of 128 L. How much milk did he carry? ____

12. Round 675 756 038 to the nearest million. ____
13. Use < (less than) or > (greater than) in:
    a 82 576 354 ____ 82 586 739
    b 98 735 089 ____ 98 735 900
14. From **A**, what is:
    a north-east? ____
    b south-west? ____

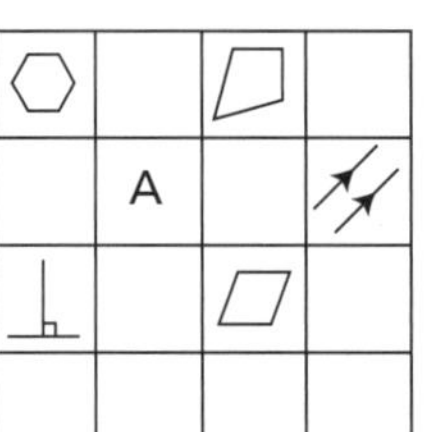


15. How many:
    a months in 9 years? ____
    b days in 11 weeks? ____
    c months in 5 centuries? ____
16. a 6 kilograms = ____ grams
    b 14 litres = ____ millilitres
    c 97 centimetres = ____ millimetres
    d 18 kilometres = ____ metres
    e 6·7 metres = ____ centimetres
    f 6 tonnes = ____ kilograms

## 12:2 out of 12

1. 560 − 24 ____
2. 489 − 49 ____
3. 457 + 79 ____
4. $\frac{5}{6} + \frac{1}{3}$ ____
5. $\begin{array}{r} 8576 \\ -\,2954 \\ \hline \end{array}$
6. 0·7 × 1000 ____
7. Days in 9 weeks. ____
8. Double 567. ____
9. Halve 878. ____
10. $\begin{array}{r} 2098 \\ \times \quad 7 \\ \hline \end{array}$
11. Each grid square is 400 km by 400 km.

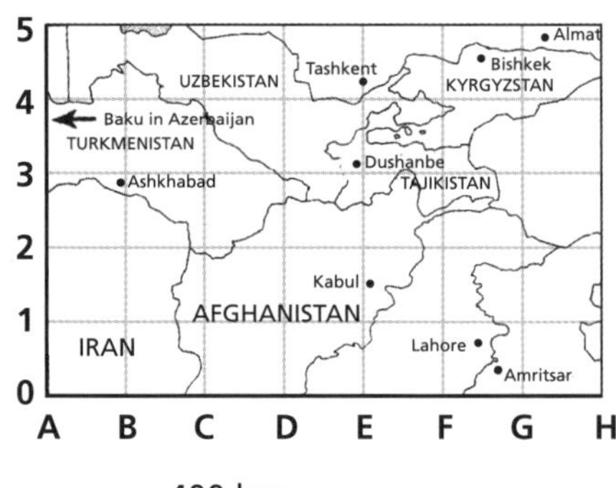

What country has coordinates of:
a G4 ____
b D1 ____
What coordinates would be used for:
c Dushanbe? ____
d Ashkhabad? ____
e What is the distance (to the nearest 100 km) between Kabul and Amritsar? ____
What city is about:
f 800 km north-east of Dushanbe? ____
g 200 km south-east of Lahore? ____

12. 100 − 9 − 9 − 9 − 9 − 9 − 9 ____

÷ Tables

| | ÷ 5 | |
|---|---|---|
| 30 | 20 | 40 |
| 10 | | 50 |
| 5 | | 25 |
| 35 | 15 | 45 |

| | ÷ 10 | |
|---|---|---|
| 80 | 50 | 20 |
| 30 | | 70 |
| 100 | | 40 |
| 60 | 90 | 10 |

To divide by 5 you can divide by 10 and double your answer.

## 12:3 out of 9

1. $4\overline{)396}$ 2. $8\overline{)256}$ 3. $10\overline{)430}$

4. Josie replaced sets of horseshoes for 253 horses this year. How many shoes did she replace? ______

5.

| h | i | j |
|---|---|---|
| k | l | m |
| n | o | p |

a What letter is north of 'o' and east of 'h'? ______

b What letter is south of 'h' and west of 'p'? ______

6. I bought eight 336 mL containers of kombucha. How many litres did I buy? ______

7. a $\frac{1}{3}$ of 15 ______

b $\frac{2}{3}$ of 15 ______

c $\frac{1}{5}$ of 15 ______

d $\frac{4}{5}$ of 15 ______

8. Plot these points and join them in the order given. **3A**, **4D**, **3E**, **2D**, **3A**

a What shape have you drawn? ______

b The diagonals intersect at: ______

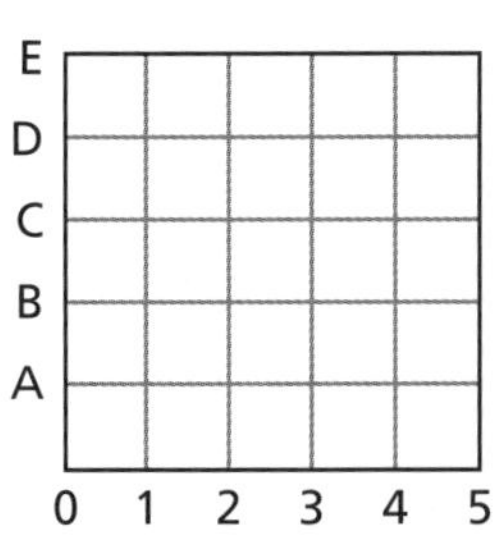


9. Each day I ride 1270 m to get to school. How many kilometres do I ride to school and back in 1 school week? ______

## 12:4 out of 6

Extension

1. 6 rolls of silk each contain a 24 m length of silk, while 2 other rolls each contain 86 m. What is the total length of silk? ______

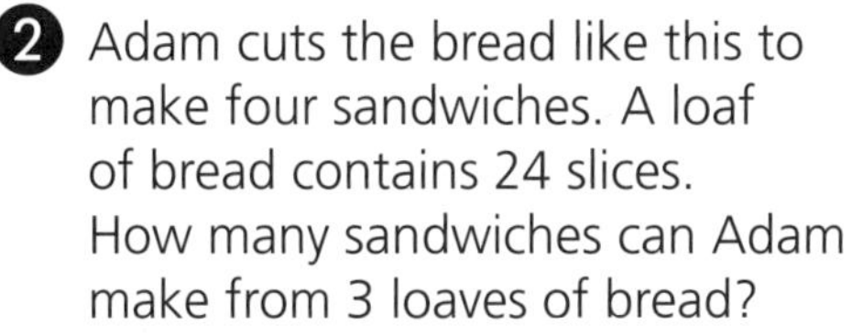

2. Adam cuts the bread like this to make four sandwiches. A loaf of bread contains 24 slices. How many sandwiches can Adam make from 3 loaves of bread? ______

3. How many zeros has the product of 3 thousand and 16 thousand? ______

4. Find the estimate by rounding each number to the nearest 10.

a $21 \times 99$ ______

b $48 \times 121$ ______

c $504 \times 21$ ______

5. Mia rode 3546 m four times this week and Jiyu rode 3157 m five times this week.

a Who rode the furthest? ______

b What was the difference between the total distances? ______

6. a $128 \div$ ______ $= 32$ b $586 -$ ______ $= 259$

**Challenge**

*Write number sentences that are equal to 24.*

______

______

______

______

______

**Heather is the youngest person in her family. This drawing has a scale of 1:50. Measure the heights in this picture using millimetres, and then find, correct to the nearest 5 cm, the real height of:**

**a** Heather ______ **b** her mother ______

**c** her brother ______ **d** her father ______

**e** Heather's dog, Shasta ______

 • *AUSTRALIAN SIGNPOST MATHS 6 MENTALS* • ISBN 978 0 6557 0886 5

## 13:1 ☐ out of 13

1. $4 \times 600$ ______
2. $800 \times 5$ ______
3. $\frac{4}{10} + \frac{1}{10}$ ______
4. $647 - 29$ ______
5. $\begin{array}{r} 76590 \\ \times \quad 2 \\ \hline \end{array}$
6. $3 \times 5 + 199$ ______
7. $(3 + 4) \times 8$ ______
8. $298 - 32$ ______
9. $21 \div 3 + 89$ ______
10. $\begin{array}{r} 95624 \\ \times \quad 4 \\ \hline \end{array}$
11. Andrew worked 5 days each week for 67 weeks. For how many days did he work? ______
12. On Makayla's first day at her new school, she studies the map at the front gate. Fill in the distance and directions using the scale.
    - **a** She walks ______ m south to order her lunch.
    - **b** She then walks 15 m ______ to the office.
    - **c** She then walks 15 m in a ______ direction to rest under a shady tree.
    - **d** If she then walks south to her classroom, what year is she in? ______
13. To square a number, multiply it by i______.

## 13:2 ☐ out of 15

1. $8 \times 25$ ______
2. $0{\cdot}6 \times 100$ ______
3. $4 - \frac{3}{4}$ ______
4. $17 - 72 \div 9$ ______
5. $\begin{array}{r} 46388 \\ + \quad 3 \\ \hline \end{array}$
6. Multiply 19 by 7. ______
7. $\frac{7}{8} - \frac{1}{4}$ ______
8. $6 \times (73 - 43)$ ______
9. $72 \div 9 \times 14$ ______
10. $\begin{array}{r} 80478 \\ \times \quad 7 \\ \hline \end{array}$
11. A photographer allows 10 minutes to take one family portrait. How many families can be photographed from 9am to 12:30pm? ______
12. A bus travels east for 4 blocks, north for 2 blocks, west for 2 blocks, then stops.

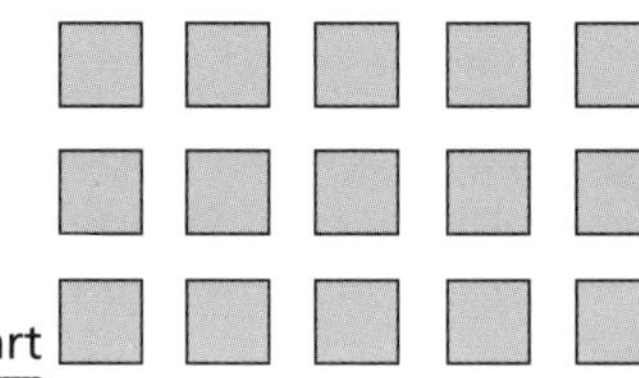


    - **a** What direction is it now facing? ______
    - **b** What direction is it now from the starting point? ______
13. If I blink 16000 times a day, estimate how many times I blink in a week. ______
14. The size of the missing angle.
    - **a** ______
    - **b** ______

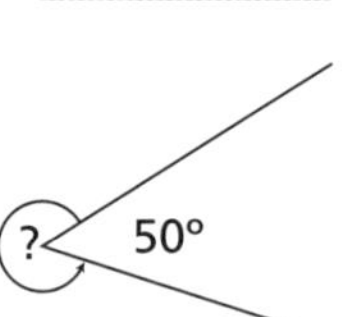


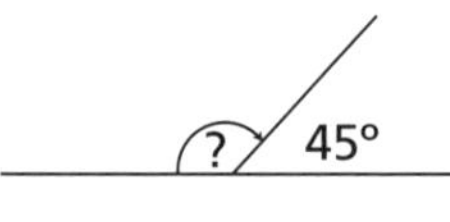


15. **a** $1000 - 20 \times (5 \times 3)$ ______
    **b** $(49 + 36) \div (45 \div 9)$ ______

Tables ÷

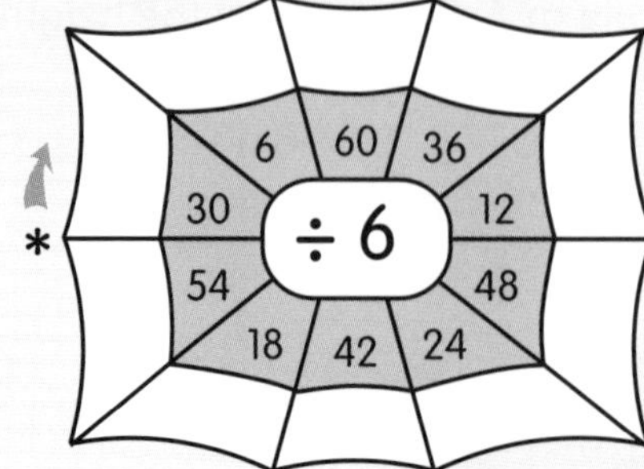

☐ × 3 = 24 ☐ = ______
☐ × 6 = 48 ☐ = ______


 ISBN 978 0 6557 0886 5

## 13:3  out of 11

1. $8\overline{)586}$  2. $5\overline{)947}$  3. $9\overline{)476}$

4. a If I need to drink 2200 mL a day, how many litres should I drink in a week? ______
   b If I only drank 12 800 mL this week, how much more did I need? ______

5. 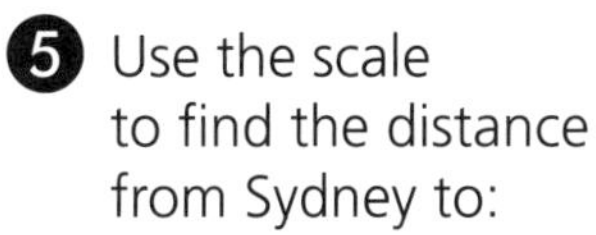 Use the scale to find the distance from Sydney to:
   a Hobart  ______
   b Perth  ______

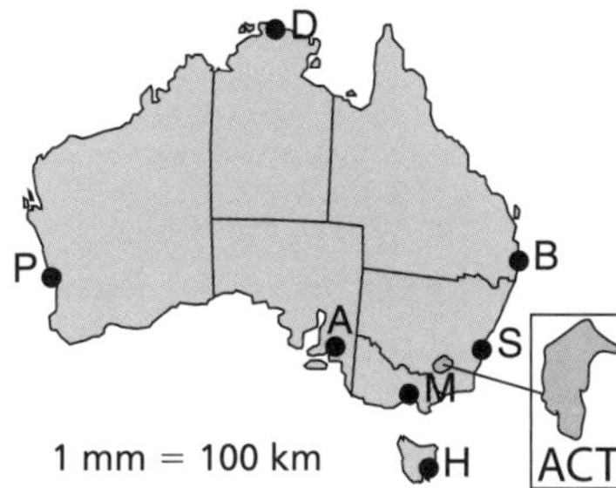


   c List the states and territories shown on this map in order of area size.
   ACT, ______

6. Adua drives 1·25 km on a sealed road and 1·25 km on a dirt road. How far has she driven altogether? ______

7. What is the size of angle **A**? ______

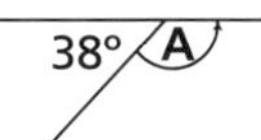


8. 48 grapes are shared fairly among 4 children. How many grapes were given to each child? ______

9. $8000 is shared by 7 adults. How much does each receive? (Round off correct to the nearest 5 cents.) ______

10. 700 + 90 + 3 + 0·06 ______

11. Does 54 354 767 round off to 54 000 000 or 55 000 000. ______

## 13:4 Extension out of 7

1. If I take about 22 500 breaths a day, how many breaths less than 1 million would I take in:
   a a week? ______
   b April? ______

2. 3 lunches cost $8.70. How much would:
   a 12 lunches cost? ______
   b 1 lunch cost? ______

3. I keep rabbits and birds. There are 12 heads and 32 feet. How many birds do I have? ______

4. A man, born in 210 BCE, died 35 years later. His grandson was born 19 years after he died.
   His grandson was born in ______.

5. If $10^3 = 1000$ and $10^2 = 100$ and $10^1 = 10$, what is the value of $10^0$? ______

6. If I blink 16 000 times a day, estimate how many times I blink in an hour. ______

7. We had 68 more red scrunchies than green scrunchies. After giving away 56, we had 356 left. How many red scrunchies did we start with altogether? ______

**Challenge**

*Write numbers sentences that are equal to 100.*

______
______
______
______

**× Tables**

×6: 7, 3, 8, 10, 6, 4, 9, 5, 0, 2

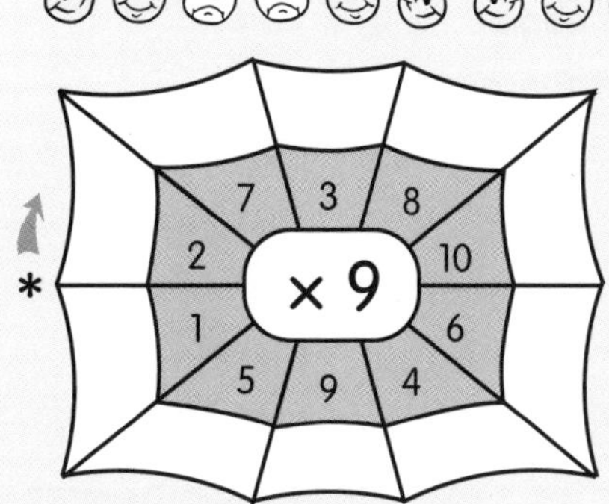


The digits in each number add up to 9.

| 9 | 18 | 27 | 36 | 45 |
|---|---|---|---|---|
| 54 | 63 | 72 | 81 | 90 |

## 14:1 — out of 17

1. 300 − 43 ______
2. $\frac{4}{6} + \frac{1}{6}$ ______
3. 849 − 427 ______
4. 0·5 × 10 ______
5. $\begin{array}{r} 2647 \\ +\ 5638 \\ \hline \end{array}$
6. 8 × (56 − 52) ______
7. Multiply 3 by 60. ______
8. 45 + 36 = ______ + 40
9. 76 − 23 = ______ − 20
10. $\begin{array}{r} 3750 \\ \times\ \ \ \ 4 \\ \hline \end{array}$
11. **a** $4\overline{)49}$ r  **b** $7\overline{)62}$ r
12. **a** 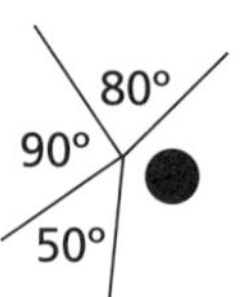



**b**

135°
45°
55°

● = ______  ▲ = ______

13. **a** In which direction does a compass needle always point? ______

**b** The angle between the compass directions, South and East? ______

14. Four friends shared $6574. How much did each receive? ______
15. A bag of marbles was dropped and 70 marbles were lost. If 297 remained, how many marbles were in the bag before it was dropped? ______
16. A temperature of 2°C dropped 10 degrees. What was it then? ______
17. The value of the 6 in 23 640 878 is ______

## 14:2 — out of 18

1. 7 + 5 × 62 ______
2. 6 × 48 ______
3. $7 - \frac{1}{3}$ ______
4. (3 + 6) × 9 ______
5. $\begin{array}{r} 8465 \\ +\ 2859 \\ \hline \end{array}$
6. $9^2 \times 8$ ______
7. $\frac{5}{6} + \frac{2}{3}$ ______
8. 845 + 298 ______
9. 345 + 456 = ______ + 450
10. $\begin{array}{r} 97860 \\ \times\ \ \ \ \ 7 \\ \hline \end{array}$
11. **a** $5\overline{)27}$ r  **b** $4\overline{)27}$ r
12. I shared 5684 mL equally into 7 drink bottles. How much did I put in each bottle? ______
13. Write as a numeral four million four thousand and forty-seven. ______
14. What size is the missing angle? (? 35°) ______
15. I pay an average of $3569 in rates for each year. What is the total cost of rates over 5 years? ______
16. Complete the labels.

**a**

______ after ______

**b**

______ to ______

17. Write this fraction:

**a** as a mixed number $\square\,\frac{\square}{\square}$

**b** as an improper fraction $\frac{\square}{\square}$

18. How many thousands in a million? ______

Turn to ID card C on page 8.

Give the answers for these numbers.

| | | | |
|---|---|---|---|
| (15) | ______ shapes | (16) | ______ shapes |
| (17) | ______ | (18) | axis of ______ |
| (19) | ______ of symmetry | (20) | ______ |
| (21) | ______ | (22) | ______ |
| (23) | ______ | (26) | ______ |

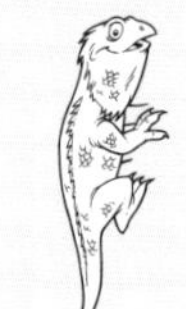

 ISBN 978 0 6557 0886 5

## 14:3 ☐ out of 14

1. $4\overline{)291}$
2. $7\overline{)704}$
3. $8\overline{)302}$
4. $7\overline{)927}$
5. $4\overline{)285}$
6. $8\overline{)809}$
7. How many seeds are in a packet if half are used to plant 5 rows of 9 seeds? ______
8. a 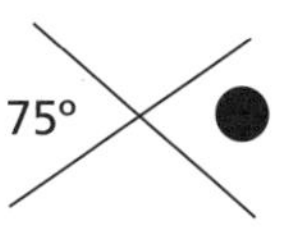

b 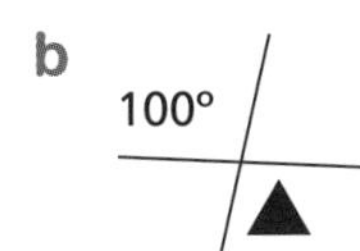

● = ______  ▲ = ______
9. If I had $35, how many of these books could I buy? ______

10. What is the value of the 4 in 6·724? ______
11. What is:
a one third of 12? ______
b one sixth of 12? ______

12. Write the numeral for two hundred and forty-seven thousand, three hundred and sixty-eight. ______
13. Write in ascending order:
1 277 456 1 097 365 1 138 000
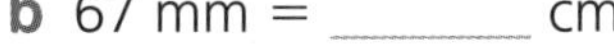
14. a 6·8 kg = ______ g
b 67 mm = ______ cm

## 14:4 Extension ☐ out of 5

1. A girl was born in 5 BCE and died 89 years later. Her great grandchild was born 16 years after she died. In what year was the grandchild born? (There is no 'zero' as a year date.) ______
2. The angle between the compass directions:
a North and North-east? ______
b South-west and North? ______

3. A plane ticket cost $2867 to India and $1878 for the return flight.
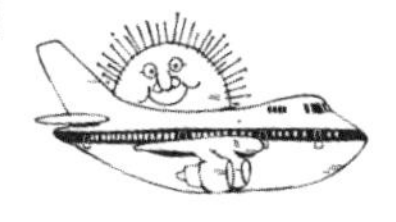
a What is the cost of the trip for 18 adults? ______
b A ticket for a child cost $1745 to India and $1517 return. If 12 children joined the adults, what would be the total cost of the adult and children's tickets? ______
4. Insert grouping symbols to make this number sentence true.
$18 - 5 \times 3 = 39$
5. An octagon with the head of a tiger inside it represents 128 tigers. What number is represented below?

______

### Challenge

*Write facts about the number 78 504 536.*

______

Turn to ID card C on page 8.
Give the answers for these numbers.

(5) ______ (6) ______
(7) ______ (8) ______
(9) ______ (10) ______
(11) ______ (12) ______
(13) ______ (14) ______

*Make a study card: Put questions on one side and answers on the other.*

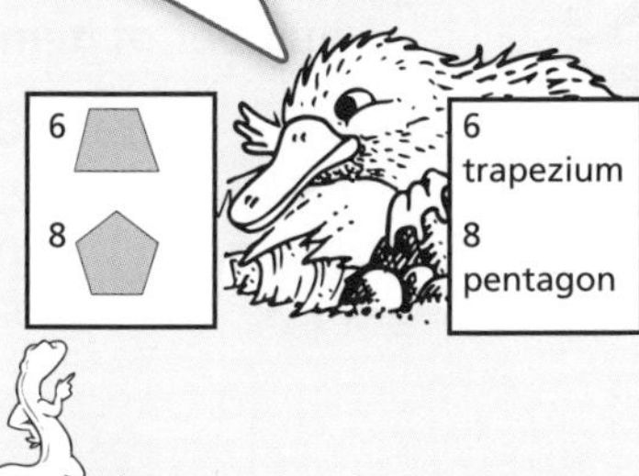

## 15:1   out of 14

1. 8 × 12 ______
2. 700 × 4 ______
3. 0·3 × 100 ______
4. 15 ÷ 3 + 9 ______
5. $8354 + $1097
6. $4.80 − 23c ______
7. $10 minus $5.20 ______
8. $3.45 − ______ = $2.85
9. $\frac{1}{4} + \frac{3}{4}$ ______
10. $1087 × 5

11. How big is the acute angle? ______

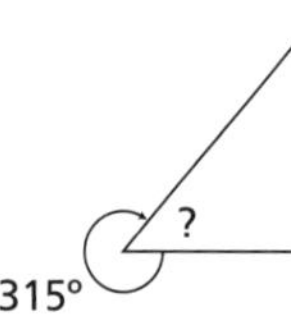

315° ?

12. How much would each person receive if 4 students equally shared 5 bananas? (Write your answer as a mixed number.) ______
13. These are the results of our soccer matches.

| | | | | | | | Totals |
|---|---|---|---|---|---|---|---|
| Our team | 5 | 4 | 2 | 0 | 8 | 7 | |
| Our opposition | 2 | 0 | 5 | 2 | 2 | 3 | |

a Write the total goals for each team.
b What was our team's average goals per game? ______
c What was the average number of goals scored by our opposition? ______
d How many games did our team win? ______

14. Write the letter used to name the point:
a (0, 3) ______
b (2, 1) ______
Write the coordinates for:
c K ______
d N ______

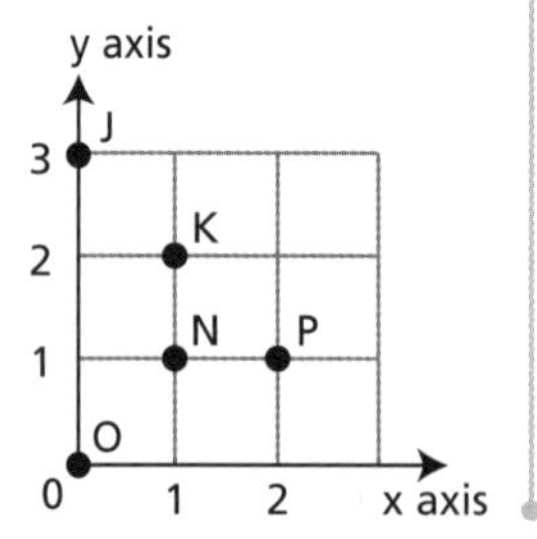


## 15:2   out of 16

1. 5 × (46 + 56) ______
2. 354 − 276 ______
3. $\frac{4}{8} - \frac{1}{2}$ ______
4. $68 × 7 ______
5. 7483 + 1987
6. ______ + 78 = 145
7. 82 + ______ = 245
8. 200 − ______ = 63
9. 75 − 43 = ______ −40
10. 3649 × 8

11. What is the average of these lengths? 67 cm, 82 cm, 94 cm, 45 cm ______
12. The size of angle **B**. ______

120° B

13. I had 4 cucumbers. The length of each was 355 mm, 376 mm, 329 mm and 341 mm. What was the average length in centimetres? ______
14. One watermelon cost $16. How many could be bought for $100? ______

15. a

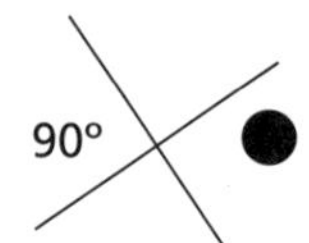

90° ●

● = ______

b 135° ▲

▲ = ______

16. Write the letter used to name the point:
a (1, 6) ______
b (6, 4) ______
Write the coordinates for:
c C ______
d B ______

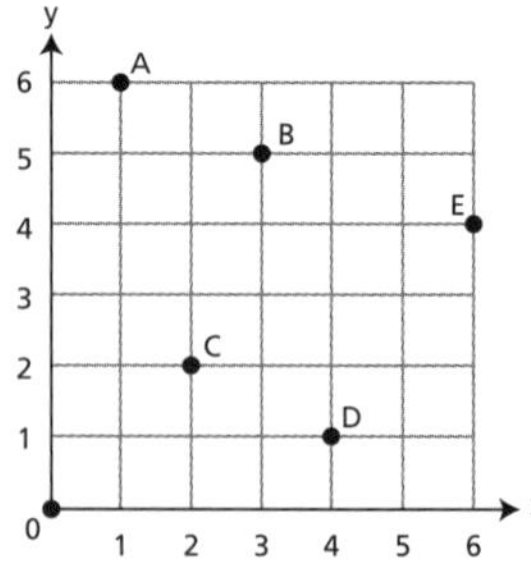

**Average**

$= \frac{\textbf{sum of the items}}{\textbf{number of items}}$

Find the average of 8, 7, 11, 27 and 12.
= (8 + 7 + 11 + 27 + 12) ÷ 5
= 13

Find the average of:

**a** 6 m, 15 m, 8 m and 7 m ______
**b** 71 g, 132 g and 88 g ______
**c** 463, 684 and 671 ______
**d** 7, 9, 1, 4 and 9 ______
**e** 81, 43, 70 and 58 ______

Add the numbers, then divide.

 • *AUSTRALIAN SIGNPOST MATHS 6 MENTALS* • ISBN 978 0 6557 0886 5

## 15:3   out of 9

1. Write the remainder as a fraction.

   a $5\overline{)43}$    b $8\overline{)34}$    c $6\overline{)835}$

   d $4\overline{)782}$    e $9\overline{)546}$    f $10\overline{)736}$

2. An angle of 360° is a ______.

3. How much would each person receive if 6 students equally shared 8 apples? (Write your answer as a mixed number.) ______

4. The size of each missing angle.

   a
   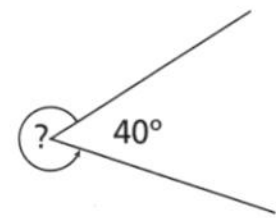
   

   b
   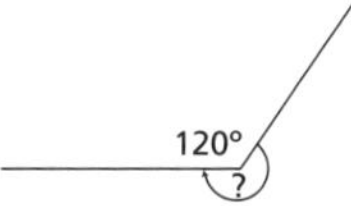
   

5. At an average speed of 80 km/h, how far would I travel in 3 hours? ______

6. Write the letter found at:

   a (0, 3) ______

   b (3, − 4) ______

   Write the coordinates for:

   c C ______

   d G ______

   e F ______

   f A ______

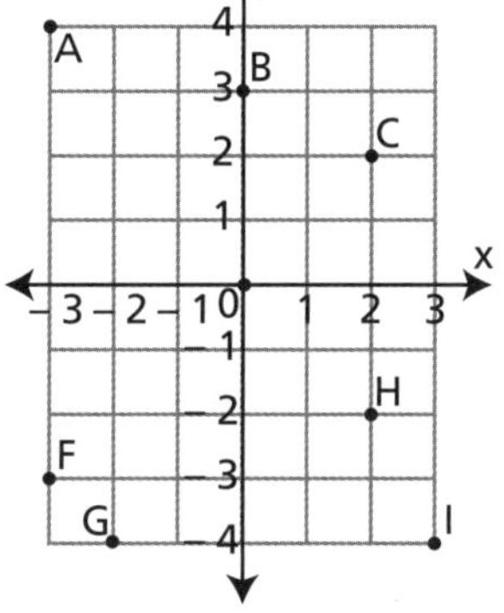


7. $(1 \times 10^4) + (8 \times 10^3) + (9 \times 10^2) + (9 \times 10^1) + 5$

   = ______

8. Write as an improper fraction:

   a $2\frac{1}{4}$ ______    b $3\frac{3}{5}$ ______

9. The value of 5 in 98·895. ______

## 15:4   out of 6

**Extension**

1. Number line: 6, A, B, C, 8, D

   Give the approximate value at the letter:

   **a A** ______    **b B** ______

   **c C** ______    **d D** ______

2. A mug contained 235 mL of water. A glass held three times as much and a bucket contained 6·54 L more than the glass. What was the total capacity of the containers? ______

3. Tickets to the show cost $169 per adult and $138.50 per child. Find the cost for our family if there are 2 adults and 6 children. ______

4. What must we add to $771 952·38 to make $900 000? ______

5. Which factor of 64 can be added to 69 to make a total of 101? ______

6. My bench top has a sink and hotplates on it.The bench top is 3·9 m by 90 cm. The sink is 85 cm by 51 cm and the hot plates cover an area 930 mm by 450 mm. How much bench space is left on the bench top? ______

**Challenge**

*List groups of numbers and record the average of each group.*

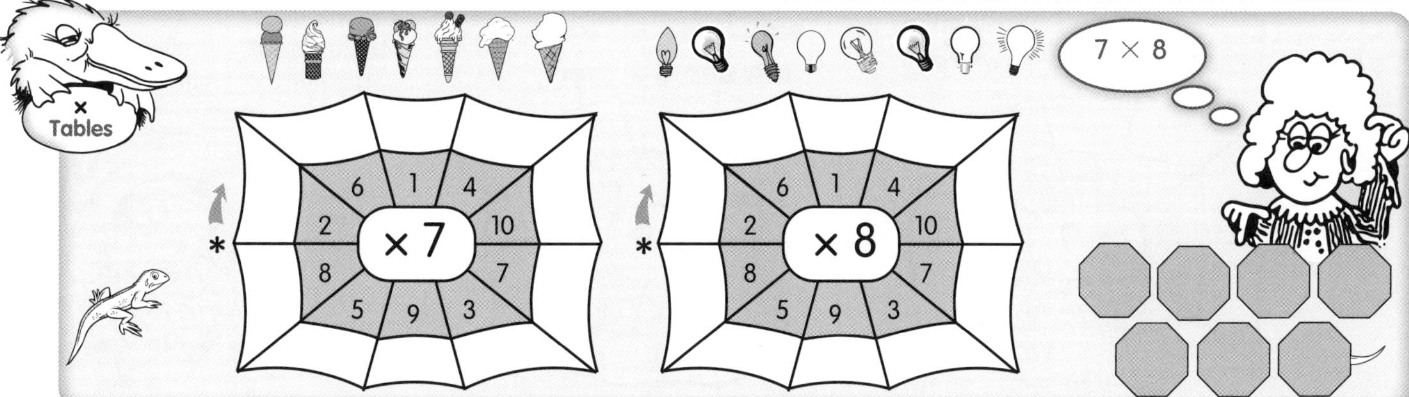


 • *AUSTRALIAN SIGNPOST MATHS 6 MENTALS* • ISBN 978 0 6557 0886 5

## 16:1 out of 14

1. $7 \times 80$
2. $42 \div 7$
3. $3 \times 7 - 8$
4. $0{\cdot}6 \times 100$
5. $\begin{array}{r} 57487 \\ +\ 20729 \\ \hline \end{array}$
6. Triple 29.
7. $\frac{1}{2}$ of 86
8. $45 + 298$
9. $\frac{6}{2} + \frac{1}{2}$
10. $\begin{array}{r} 30586 \\ +\ 18474 \\ \hline \end{array}$
11. How much would each receive if six students equally shared 10 cucumbers? (Write your answer as a mixed number.)
12. At an average speed of 60 km/h, how far would I travel in 2 hours?
13. **a** Draw a dot at (2, 3), (−1, 3), (−2, −2) and (1, −2).
    **b** Join the dots to make a shape.
    **c** What shape did you draw?

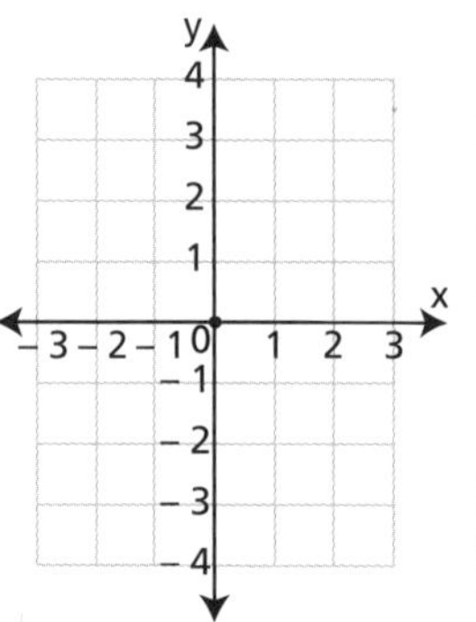


14. **a** Which solid does this model represent?
    **b** What is the cross-section of this solid?
    **c** List all vertical lines.
    **d** List all horizontal lines.

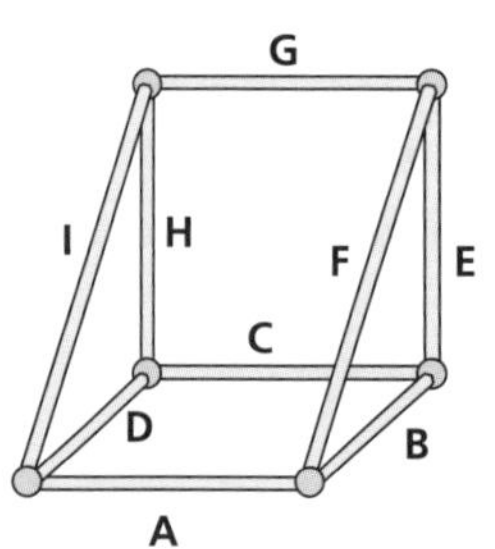

## 16:2 out of 16

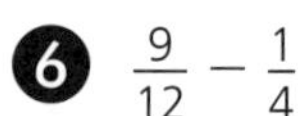

1. $763 + 157$
2. $465 - 156$
3. $56 \div 8 + 9$
4. $42 \div 7 - 6$
5. $\begin{array}{r} 419676 \\ +\ 36849 \\ \hline \end{array}$
6. $\frac{9}{12} - \frac{1}{4}$
7. $365 + 927$
8. $\frac{1}{8}$ of 24
9. $84 + 43 = 87 +$ ____
10. $\begin{array}{r} 435790 \\ -\ 56841 \\ \hline \end{array}$
11. How far can a kangaroo travel if it moves at an average speed of 46 km/h for half an hour?

12. $(9 \times 10^4) + (2 \times 10^3) + (7 \times 10^2) + (4 \times 10^1) + 6$ = ____
13. **a** Draw a dot at (1, 1), (3, −3), (−3, −3) and (−1, 1).
    **b** Join the dots to make a shape.
    **c** What shape did you draw?

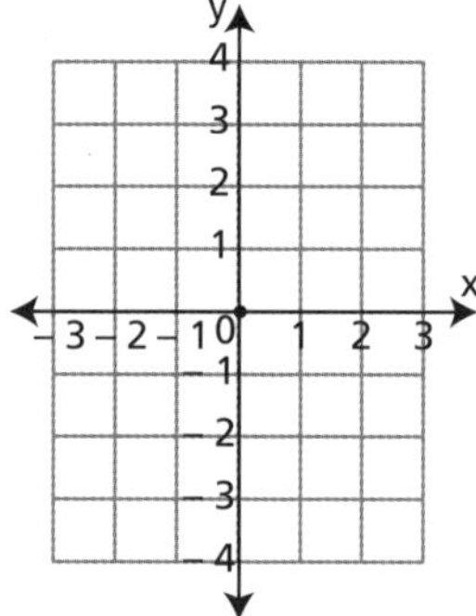


14. A supermarket purchased 465 475 apples this month and sold 287 365 in the first 10 days.
    **a** How many have not been sold?
    **b** Is it likely that they will all be sold by the end of the month?
15. These pears weighed 197 g, 218 g and 188 g. What was the average mass?

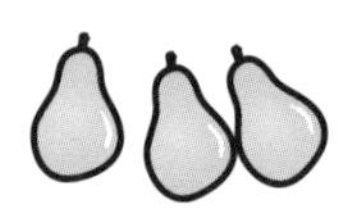

16. Name solids that have a square cross-section.

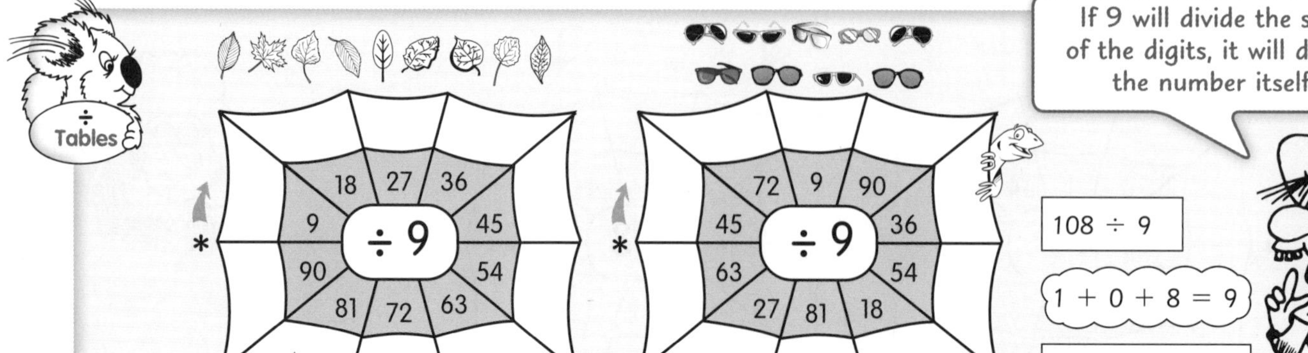

## 16:3 — out of 10

1. Write each answer as a mixed number.

   a 7)57  b 5)94  c 3)695

   d 6)693  e 8)639  f 10)392

2. 
```
  36705
  14054
+ 23109
```

3. 
```
  73000
  14000
   3100
+  1053
```

4. How far would a dog travel if it ran at an average speed of 12 km/h for 10 minutes? ______

5. The size of:

   a angle **A** ______

   b angle **B** ______

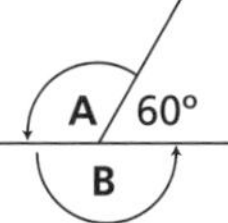


6. How much would each person receive if 3 students equally shared 11 slices of bread? ______

7. I had 4 bananas with a mass of 105 g, 129 g, 116 g and 109 g. What is the average mass? ______

8. Sophie bought a house for $568 000 and sold it for $934 550. What was the difference between the purchase and sale price? ______

9. There were 46, 52 and 73 fish in each of three fish tanks. What was the average number of fish in the tanks? ______

10. A cube has ______ faces and ______ edges.

## 16:4 — Extension — out of 6

1. Toothpicks and Blu Tack were used to make triangular prisms. What is the greatest number of separate triangular prisms you could create with:

   a 19 toothpicks? ______

   b 27 toothpicks? ______

2. One plane flew 13 km per minute for 6 hours. Another plane flew 860 km/h for 6 hours. What was the difference between the total distance they flew? ______

3. I used square pieces of wood 4 cm long to construct open cubes. How many open cubes could I make using 79 of the square pieces? ______

4. If # stands for 7356 and ~ stands for 193, what is 2 × (8000 − # − ~) − (4 × ~). ______

5. a 266 × 34 ______

   b 675 × 26 ______

6. The teachers set out 29 rows of 34 chairs for the assembly. If 896 people sat in the chairs during the assembly, how many chairs were not used? ______

### Challenge

*Make your own rule to complete each table.*

Rule = × ______ + ______

| 1st number | 1 | 2 | 3 | 4 | 5 | 6 | 7 |
|---|---|---|---|---|---|---|---|
| 2nd number | | | | | | | |

Rule = × ______ − ______

| 1st number | 1 | 2 | 3 | 4 | 5 | 6 | 7 |
|---|---|---|---|---|---|---|---|
| 2nd number | | | | | | | |

### Profit and loss

When we sell something for more than we paid for it, we make a **profit**. When we sell something for less than we paid for it, we make a **loss**.

Give the profit or loss, if an item:

**a** bought for $137.50 is then sold for $150. Profit of ______

**b** bought for $18 250 is then sold for $10 850. ______ of ______

**c** bought for $9380 is then sold for $16 250. ______ of ______

 *AUSTRALIAN SIGNPOST MATHS 6 MENTALS* • ISBN 978 0 6557 0886 5

## 17:1 ☐ out of 19

1. 600 − 38 ______
2. 200 − 152 ______
3. 36 ÷ 6 + 3 ______
4. 27 ÷ 9 + 6 ______
5. $\begin{array}{r} 507484 \\ +\ 28018 \\ \hline \end{array}$
6. $\frac{1}{2}$ of 362 ______
7. 0·6 × 100 ______
8. Multiply 90 by 7. ______
9. Divide 100 by 4. ______
10. $\begin{array}{r} 809230 \\ -\ 28560 \\ \hline \end{array}$
11. Three friends shared $4656. How much did each receive? ______
12. This is the net of a ______________________.
13. What is the cross-section of a square pyramid parallel to the base? ______
14. What is the probability of spinning G:
    - a as a fraction? ______
    - b as a decimal? ______
    - c as a percentage? ______

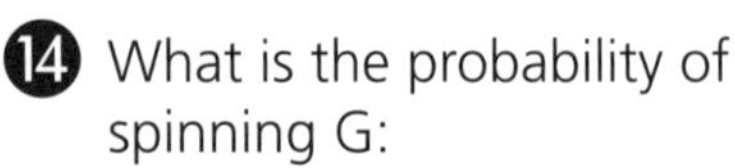

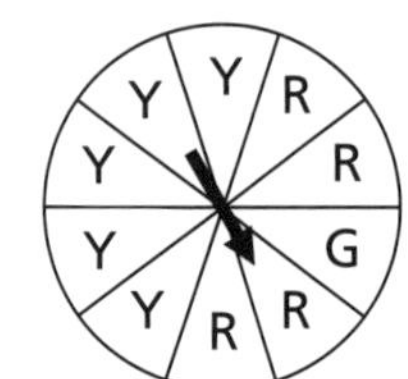


15. Jessica wins if a Y is spun, Felicity wins if an R is spun and Lachlan wins if a G is spun.
    - a Is this a fair game? ______
    - b Who has the best chance of winning? ______
16. 0·8, 1·0, 1·2, ______, ______, ______, ______
17. Write $3\frac{4}{5}$ as an improper fraction. ______
18. a How many halves in $2\frac{1}{2}$? ______
    b How many quarters in 3? ______
19. The value of 6 in 857·786. ______

## 17:2 ☐ out of 17

1. 5198 − 36 ______
2. 1000 − 79 ______
3. 800 − 156 ______
4. 560 ÷ 7 ______
5. $\begin{array}{r} 800000 \\ -\ 13648 \\ \hline \end{array}$
6. $\frac{1}{5}$ of 45 ______
7. 50% of $62 ______
8. 0·8 as a fraction ______
9. 0·65 as a percentage ______
10. $\begin{array}{r} 900000 \\ -\ 26478 \\ \hline \end{array}$
11. I scored 91% in Maths, 82% in English, 82% in Science. What was the average of my 3 scores? ______
12. A plane can fly at 14 km per minute. How far can it fly in one hour? ______
13. My odometer now shows 39 873 km. How far must I travel until my next service which is at 50 000 km? ______
14. a Complete the drawing of this shape.

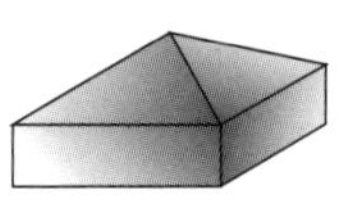

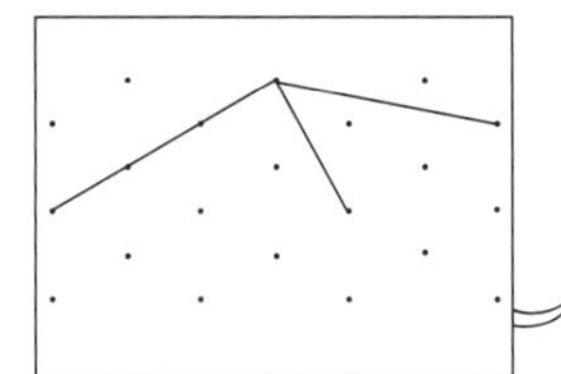

This shape has:

b ____ vertices c ____ faces

d Describe the cross-sections of this shape that are parallel to the base. ______

15. This is the net of a ______________________

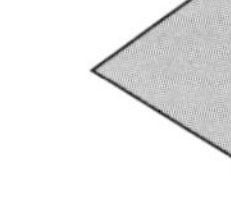

16. What is the cross-section of a triangular prism parallel to the base? ______
17. The value of 9 in 3546·893 ______

______________________
______________________
______________________
______________________
______________________


 • ISBN 978 0 6557 0886 5

## 17:3 out of 11

**1**

```
   576476
    64769
  3568401
+    3567
```

**2**

```
  900000
−  35975
```

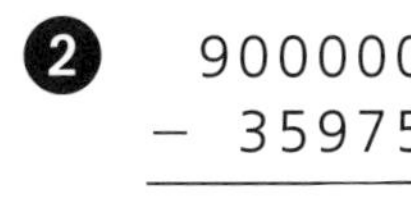

**3** The aeroplane was at a height of 10 232 m. An hour later it was at a height of 12 588 m. What was the difference between the heights? ______

**4** This is the net of a ______.

**5** What is the cross-section of a hexagonal prism, parallel to the base? ______

**6** What is the probability of spinning Y:
- **a** as a fraction? ______
- **b** as a decimal? ______
- **c** as a percentage? ______

**7** 100 tickets are sold at a raffle. Harvey bought 4. What is his chance of winning as a:
- **a** percentage? ______
- **b** fraction? ______
- **c** decimal? ______

**8** Complete each pattern.
- **a** 0·2, 0·8, 1·4, ______, ______, ______, ______
- **b** 0·3, 0·6, 0·9, ______, ______, ______, ______

**9** The value of the 9 in 4657·97. ______

**10** (4 × 900) + (3 × 600) = ______

**11**

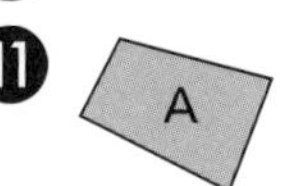

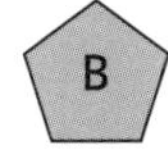

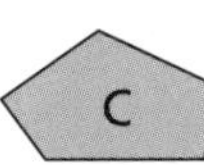

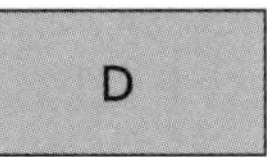

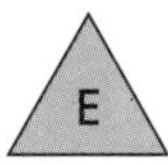

- **a** Which shapes are regular? ______
- **b** Which shapes have all angles obtuse? ______

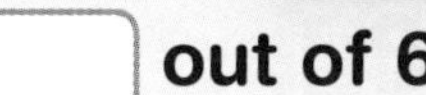


## 17:4 out of 6

**1** What is the smallest number of matchsticks needed to make:
- **a** 7 small hexagons? ______
- **b** 6 small triangles? ______

**2** The number of faces plus the number of corners minus the number of edges on 19 triangular prisms. ______

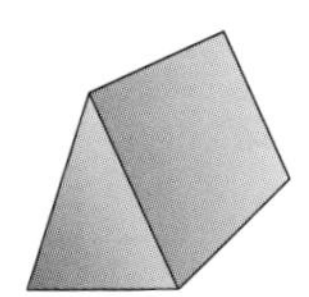

**3** I used square pieces of wood 8 cm long to construct open cubes. How many open cubes could I make using 127 of the square pieces? ______

**4** True or false?
- **a** 0·60 ≠ 0·06 ______
- **b** $0{\cdot}70 = \frac{7}{10}$ ______

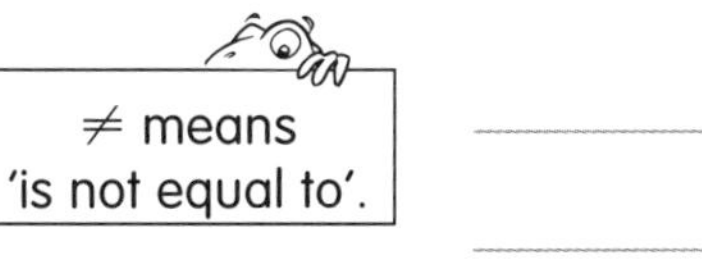


**5** I stands for 1, V stands for 5, X stands for 10, L stands for 50 and C stands for 100. Write the number for:
- **a** LXXXI ______
- **b** CLXII ______
- **c** CCXVI ______
- **d** CCVI ______

**6** If # stands for 887·978 and ~ stands for 168·99, what is 9 × (# + ~) − (2 × ~). ______

### Challenge

*Write questions that are equal to:*

**a** 987 765 + 786 543
= ______
= ______
= ______
= ______
= ______

**b** 987 675 + 235 432
= ______
= ______
= ______
= ______
= ______

**What is the probability of success?**

**Estimate the place for each letter and write it above the scale.**

**A** The next baby born in Melbourne will be a girl.
**B** It will rain tomorrow.
**C** I will go to the movies next month.
**D** I will get a 6 if I throw a dice once.
**E** Our teacher will become prime minister tomorrow.

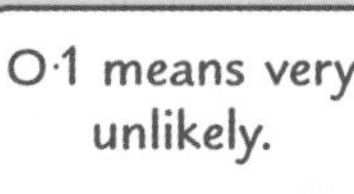


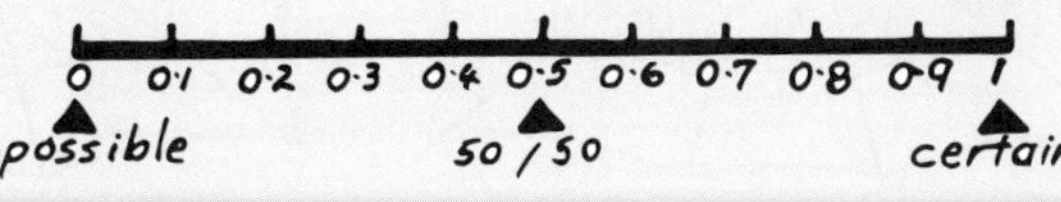

## 18:1 out of 15

1. 18 ÷ 3 ______
2. 42 ÷ 6 ______
3. 18 ÷ 2 − 7 ______
4. 0·7 × 100 ______
5. 354657 + 36509
6. 40 ÷ 10 + 76 ______
7. 36 divided by 6. ______
8. $\frac{1}{3}$ of 63 ______
9. 7 × $93 ______
10. 809780 − 47293

11. For this spinner, which colour's chance of occurring is:

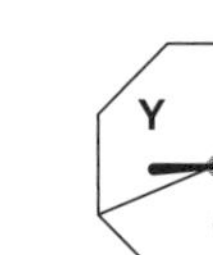


Y = yellow
B = blue
G = green

a least likely? ______
b an even chance? ______
c a 50% chance? ______
d a $\frac{1}{8}$ chance? ______

12. How many 20c coins have the same value as $5.40? ______
13. Five friends shared 3657 mL of water. How much water did each get? ______
14. Heidi began with $30 and bought these items.

Total spent = ______

Amount left = ______

15. a $\frac{4}{6} + \frac{1}{6}$ ______ b $\frac{10}{12} - \frac{5}{12}$ ______

## 18:2 out of 15

1. 820 ÷ 9 ______
2. $\frac{1}{9}$ of 45 ______
3. 0·8 × 100 ______
4. $\frac{3}{4}$ of 80 ______
5. 400000 − 89977
6. 467 + 635 = ______ + 640
7. 587 − 227 = ______ − 30
8. $645 − 80c ______
9. $56.30 − $5.85 ______
10. $8000·00 − $7987·90

11. Two heads, two tails, or a head and a tail could result from tossing two coins. What is the most likely result? ______
12. Which letter has a 25% chance that the spinner will land on it? ______

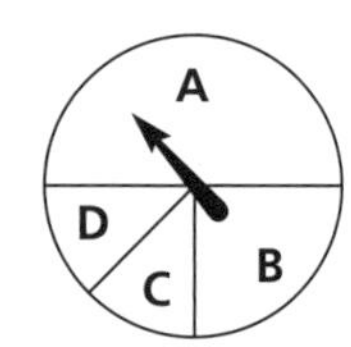


13.

| Income | Gabby | Priscilla | Ilya |
|---|---|---|---|
| Mowing lawns | $30 | 0 | $10 |
| Babysitting | 0 | $28 | $12 |
| Washing cars | $18 | $8 | $12 |
| Odd jobs | $16 | $23 | $25 |
| TOTAL | $ ______ | $ ______ | $ ______ |

a How much was earned by Gabby, Priscilla and Ilya altogether? ______
b Round this total to the nearest ten dollars. ______

14. Find the total of these amounts, rounded off to the nearest 5 cents. ______

$ 5.30
$ 9.00
$ 2.50
$ 8.26

15. 63 ÷ (16 − 7) = ______

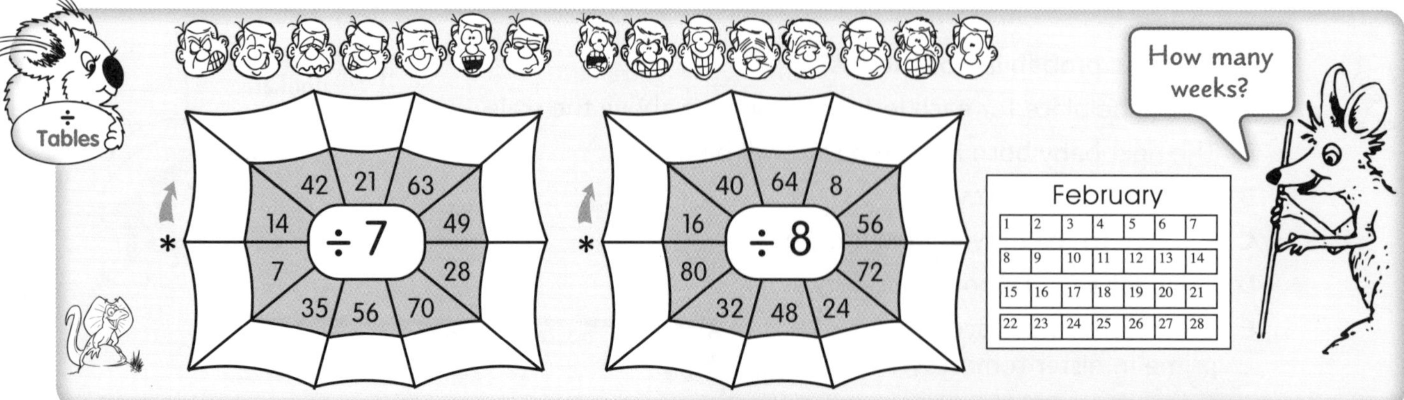




 • *AUSTRALIAN SIGNPOST MATHS 6 MENTALS* • ISBN 978 0 6557 0886 5

# Answers

## ID card answers

### ID card A

1 millimetres **2** centimetres **3** metres **4** kilometres **5** scale
**6** milligrams **7** grams **8** kilograms **9** tonnes **10** millilitres **11** litres
**12** kilolitres **13** megalitres **14** square centimetres **15** square metres
**16** 1 square metre **17** hectares **18** 10 000 $m^2$ **19** square kilometres
**20** cubic centimetres **21** 1 mL **22** cubic metres **23** seconds **24** minutes
**25** hours **26** before noon **27** after noon **28** 5:40 am **29** 19:30
**30** degrees Celsius

### ID card B

**1** integers **2** prime number **3** composite number **4** 7 **5** 2 and 4
**6** 5 **7** remainder **8** square numbers **9** digits **10** 4 ones, 3 tenths, 9 hundredths, 5 thousandths **11** 19 out of 100 **12** vertical line
**13** horizontal line **14** interval **15** ray **16** parallel lines
**17** perpendicular lines **18** corner **19** arm **20** acute **21** right angle
**22** obtuse angle **23** straight angle **24** reflex angle **25** revolution
**26** 90° **27** 180° **28** 360° **29** A = C, B = D **30** protractor

### ID card C

**1** circle **2** oval **3** triangle **4** square **5** rectangle **6** rhombus
**7** parallelogram **8** trapezium **9** kite **10** quadrilaterals **11** pentagons
**12** hexagons **13** octagon **14** decagon **15** regular shapes
**16** irregular shapes **17** diagonals **18** axis of symmetry
**19** axes of symmetry **20** rotational symmetry **21** reflection
**22** translation **23** rotation **24** coordinates **25** number plane
**26** tessellation **27** P = 2W + 2L (or P = W + W + L + L)
**28** A = L × W **29** A = $s^2$ **30** rectangle

### ID card D

**1** face **2** corner or vertex **3** edge **4** cube **5** rectangular prism
**6** triangular prism **7** hexagonal prism **8** triangular pyramid
**9** square pyramid **10** rectangular pyramid **11** base
**12** cylinder **13** cone **14** sphere **15** net of a cube **16** net of a cone
**17** net of a cylinder **18** net of a square pyramid **19** cross-section
**20** picture graph **21** tally **22** column graph **23** line graph
**24** dot plot **25** sector graph **26** divided bar graph
**27** median **28** mode **29** range **30** average

## 1:1

**1** 54 **2** 9 **3** $18 **4** 4 **5** 8048 **6** 667 **7** 21 **8** 3 **9** 1
**10** 8506 **11** a $\frac{1}{6}$ b $\frac{4}{6}$ **12** 49 300 000 **13** $\frac{4}{8}$ or $\frac{1}{2}$
**14** a 48% b 29% **15** a 32, 36, 40, 44 b 24, 27, 30, 33
c 37, 42, 47, 52 **16** 12 **17** a 2 m 56 cm b 4·6 cm c 4000 g
d 4·75 kg **18** $175

## 1:2

**1** 8 **2** $42 **3** 551 **4** 43 **5** 5706 **6** 9 **7** $32
**8** 386 **9** 50 **10** 26 540 **11** 6500 **12** $\frac{8}{12} - \frac{4}{12} = \frac{4}{12}$
**13** a 0·51, 0·5, 0·49 b subtract 0·01
**14** 4 × 164 = (4 × 100) + (4 × 60) + (4 × 4) = 400 + 240 + 16 = 656
**15** $6.30 or 630c **16** a 1200 b 1200 c 1500 d 4200
**17** $5\frac{4}{6}$, $5\frac{5}{6}$, $5\frac{6}{6}$ (or 6) **18** 300 minutes **19** $\frac{9}{10}$ **20** 68%

### Activity

26, 79, 20, 44, 33, 55, 31, 42, 47, 58
28, 81, 22, 46, 35, 57, 33, 44, 49, 60
even + odd = odd
odd + odd = even

## 1:3

**1** 108 **2** 127 **3** 167 **4** 745 321 **5** 7, 14, 21, 28, 35, 42
**6** 97 774 421 **7** 40 **8** 6 **9** 36 $m^2$ **10** 40
**11** 5 square kilometres **12** 1, 2, 3, 4, 6, 12 **13** 60 squares will be coloured red, and 30 squares will be coloured blue. $\frac{90}{100}$ or $\frac{9}{10}$
**14** a 376 b 699 c 999 **15** 36 **16** a 155, 145, 135
b 40, 48, 56, 64 **17** 77 **18** 9

## 1:4

**1** 135th **2** a [6] [4] × 3 = [1] 9 2 b 3)[3] [8] [7] = 1 2 9
**3** 810 **4** a 160 b 460 c 35 460 **5** 24 **6** $\frac{15}{24}$
**7** a 7 b 9 **8** a 5 b 9

### Challenge

Answers may vary. E.g. It has 5 digits. It becomes 43·83 when rounded to the nearest hundredth.

### Activity

Answers will vary.

## 2:1

**1** 6 **2** 18 **3** 80 **4** 100 **5** 8912 **6** 222 **7** 10 **8** $28
**9** 4 **10** 7160 **11** 24 **12** 5:56 am **13** $15 will be circled.
**14** Answers will vary. **15** 8 781 344, 8 780 033, 8 768 367
**16** 46 400 **17** 35 483 **18** a 5 L b 7300 mL c 7·365 km
d 5538 mL **19** a 36, 42, 48, 54 b 42, 49, 56, 63

## 2:2

**1** 32 **2** 30 **3** 5 **4** 5 **5** 2816 **6** 109 **7** 409
**8** $72 **9** 9 **10** 11 253 **11** $\frac{1}{6}$ **12** 52 000 000 **13** $\frac{65}{100}$
**14** nine million, seven hundred and fifty-six thousand.
**15** a 49 squares will be shaded. b 51 **16** 2·63 **17** a 3 b 9
**18** a 250 cm b 25 mm **19** 24 m

### Activity

60, 56, 85, 93, 88, 71, 85, 67, 39, 72
58, 48, 63, 41, 79, 86, 78, 65, 77, 70
even − even = even
odd − even = odd

## 2:3

**1** 109 **2** 203 **3** 96 **4** 2456 **5** a 6·09 b 19·3 **6** 137
**7** 9 o'clock and 3 o'clock **8** 16 cm **9** 16:03
**10** a 1300 mL or 1 L 300 mL b 700 mL
**11** 2 hundreds, 0 tens, 1 one, 7 tenths, 9 hundredths **12** a $\frac{3}{4}$ b $\frac{1}{4}$

## 2:4

❶ a 10 b 12 ❷ 5 ❸ 9 ❹ a $870 b $1044
❺ a 435 + 297 = 432 + 300 = 732
b 824 − 392 = 822 − 390 = 432 c 725 − 409 = 726 − 410 = 316
❻ a 81 b 486

### Challenge

Answers will vary.
E.g. 30 + 6 = 36, 42 − 6 = 36, 3 × 12 = 36, etc.

### Activity

(1) face (2) corner or vertex (3) edge (12) cylinder (13) cone (14) sphere (15) net of a cube (16) net of a cone (17) net of a cylinder (18) net of a square pyramid

## 3:1

❶ 20 ❷ 15 ❸ 3 ❹ 4 ❺ 2919 ❻ 141 ❼ 864 ❽ 40
❾ $62 ❿ 16 126 ⓫ a G = 8, S = 12, B = 16 b Bronze c 8
⓬ 46 m ⓭ a C b A ⓮ 7 050 321

## 3:2

❶ 25 ❷ $496 ❸ 3 ❹ 4 ❺ 8280 ❻ 178 ❼ $56 ❽ 6
❾ 7 ❿ 50 776
⓫ a

3:04 am

b

7:22 pm

⓬ 36 ⓭ 40 ⓮ a 3:02 pm b 5:39 pm c 7:53 am d 4:28 am
⓯

| Pentagons | 1 | 2 | 3 | 4 | 5 |
|---|---|---|---|---|---|
| Sides | 5 | **10** | **15** | **20** | **25** |

⓰ 15 ⓱ 45 ⓲ 171

### Activity

8, 40, 24, 80, 56, 32, 64, 16, 72, 48
5, 25, 15, 50, 35, 20, 40, 10, 45, 30

## 3:3

❶ 166 ❷ 117 r 6 ❸ 94 ❹ 32 ❺ 14 ❻ 75 ❼ 16·25
❽ 1·2, 0·99, 0·8, 0·67 ❾ true ❿ 56
⓫

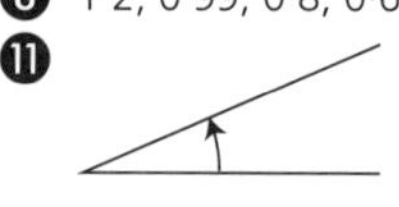

acute

obtuse

reflex

⓬ 4002·21 ⓭ 2 033 469

## 3:4

❶ 44 407 kg ❷ a 10 min b 20 min ❸ 2 ❹ 5 ❺ 5
❻ 1281

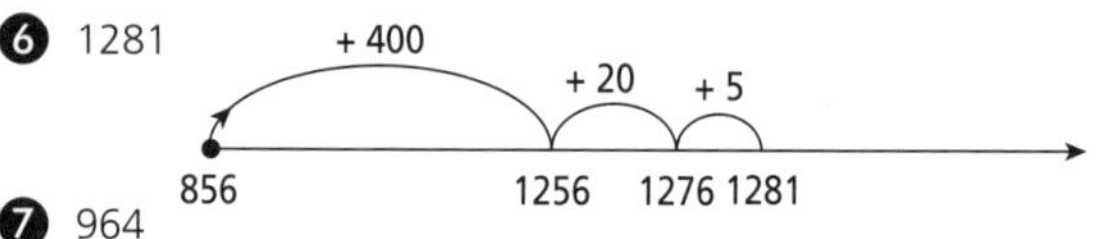


❼ 964

### Challenge

Answers may vary.
E.g. a 740 + 276, 808 + 208, etc. b 204 + 447, 300 + 351, etc.
c 3000 + 5360, 2678 + 5682, etc.

### Activity

a $37.45 b $43.90 c $30.00 d $84.65 e $117.10 f $265.15
g $630.95 h $365.30 i $289.10 j $836.35

## 4:1

❶ 16 ❷ 32 ❸ 4 ❹ 8 ❺ 8027 ❻ 796 ❼ 30 ❽ 50
❾ 846 ❿ 18 624 ⓫ 1, 3, 5, and 15. ⓬ 6, 12, 18, 24, 30, 36, 42, 48, 54, 60 ⓭ a 16 b 64 ⓮ 3 toys ⓯ a 9 b 7 ⓰ 80
⓱ 35 ⓲ 85 cm ⓳ 3 000 000 ⓴ 6 tenths or 0·6 ㉑ 0·51 ㉒ 4

## 4:2

❶ 25 ❷ 2 ❸ 9 ❹ 27 ❺ 5608 ❻ 40 ❼ 72 ❽ 240
❾ 300 ❿ 59 704 ⓫ 9 ⓬ a 7 b 7 ⓭ 7 ⓮ 2625 mL or 2·625 L
⓯ 9 ⓰ a 24 b 8 ⓱ 480 ⓲ a 1 b 2 ⓳ a 0·3 b $\frac{3}{10}$
⓴ 3, 4, and 6 will be circled.

### Activity

124, 116, 174, 190, 180, 146, 174, 138, 82, 148
208, 96, 324, 196, 212, 272, 296, 392, 228, 184

## 4:3

❶ 109 ❷ 271 ❸ 181 ❹ 73 ❺ 1, 24, 2, 12, 3, 8, 4, 6
❻ 8, 16, 24, 32, 40, 48, 56, 64, 72, 80 ❼ a 9 b 81 ❽ 6
❾ 18 ❿ a 7 b 7 ⓫ a 14 cm or 0·14 m b 96·81 kg ⓬ 9
⓭ 14, 12, 9 and 7

## 4:4

❶ 72 and 73. ❷ 115 ❸ 25 h 15 min ❹ 4 ❺ 5036 g or 5·036 kg
❻ a 30 b 36 ❼ 39

### Challenge

Answers will vary.
E.g. a 900 − 289, 614 − 3, etc. b 700 − 302, 474 − 76, etc.
c 7159 − 4000, 7659 − 4500, etc.

### Activity

762, 126, 103, 434, 169, 231, 150, 79, 768, 147
164, 933, 100, 355, 947, 854, 642, 71, 659, 778
even + even = even
odd + even = odd

## 5:1

❶ 21 ❷ $12 ❸ 3 ❹ 5 ❺ 993 ❻ 4 ❼ 4 ❽ 35 ❾ 70
❿ 7944 ⓫ 1, 12, 2, 6, 3, 4 ⓬ 3 ⓭ 6 ⓮ a 12 b 4
⓯ 31 673 316, 31 763 116, 33 761 301, 37 631 713 ⓰ a 14
b 504 c 66 ⓱ 28 ⓲ 100 ⓳ 9 ⓴ 98% ㉑ 7277


 *AUSTRALIAN SIGNPOST MATHS 6 MENTALS* • ISBN 978 0 6557 0886 5

## 5:2

**1** 54 **2** $\frac{2}{6}$ **3** \$5 **4** 9 **5** 8753 **6** 25 **7** 45 **8** \$9 **9** 900

**10** 17 670 **11** 7, 14, 21, 28, 35, 42, 49, 56, 63, 70

**12** **a** 4 groups 2 left **b** 5 groups 1 left **13** **a** 7 **b** 9 **14** 31 and 33.

**15** **a** 700 **b** 1200 **16** 420 cm or 4·2 m **17** 8

**18**

| | |
|---|---|
| 0·61 | 80% |
| 0·49 | 61% |
| 0·80 | 25% |
| 0·10 | 49% |
| 0·25 | 10% |

### Activity

**a** 20% **b** 12%

## 5:3

**1** 144 **2** 309 **3** 102 r 5 **4** 9063 **5** 32 599

**6** Books each = 5, Remainder = 2 **7** 1, 2, 3, 4, 6, 9, 12, 18, and 36.

**8** **a** 7 **b** 7 **9** **a** 15 and 60 **b** 1 and 49 **10** 15 **11** \$396

**12** **a** 4208 **b** 4734 **13** 114 + 425 = 539 or ☐ + 114 = 539

**14** 1% **15** 8 000 000

## 5:4

**1** 16 **2** $2^2 + 3^2 + 4^2$ **3** Drawing will be copied. **4** 110 **5** 40

**6** **a** yes **b** no **7** \$83.10 **8** 294

### Challenge

Answers will vary.

### Activity

| | | | | |
|---|---|---|---|---|
| | $\frac{3}{10}$ | $\frac{30}{100}$ | 0·3 | 30% |
| | $\frac{6}{10}$ | $\frac{60}{100}$ | 0·6 | 60% |

## 6:1

**1** 47 **2** 4 **3** 6 **4** 4 **5** 6342 **6** 27 **7** 33 **8** 54 **9** 47

**10** 4269 **11** 81 out of 100.

**12**

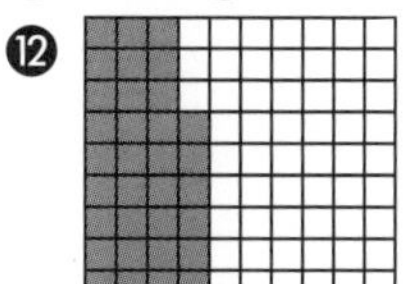

| 0·37 | $\frac{37}{100}$ | 37% |
|---|---|---|

**13** 67 747 324, 67 647 324, 67 541 234 **14** 95 747 013

**15** 35 000 000 **16** **a** 24 **b** 43 **17** rectangle **18** **a** cylinder **b** sphere **19** 18

## 6:2

**1** 37 **2** 36 **3** 8 **4** 6 **5** 1877 **6** $\frac{5}{12}$ **7** 532 **8** 162

**9** 172 **10** 14 588 **11** 600 000 **12** 76 out of 100 = $\frac{76}{100}$ = 0·76 = 76%

**13** 53 803 **14** reflection **15** 1, 42, 2, 21, 3, 14, 6, 7

**16** 73 000 000 **17** **a** 12 **b** 29 **18** A **19** rotation

### Activity

**a** 6 **b** 4 **c** 16 **d** 4 **e** 17 **f** 17 **g** 4 **h** 16 **i** 12 **j** 16 **k** 63

## 6:3

**1** 46 **2** 156 **3** 68 r 3 **4** 2137 **5** 19 045 **6** 8

**7**

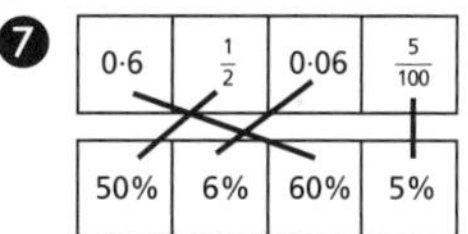

| 0·6 | $\frac{1}{2}$ | 0·06 | $\frac{5}{100}$ |
|---|---|---|---|
| 50% | 6% | 60% | 5% |

**8** **a** 35% **b** 25% **9** 56 499 **10** 89 000 000 **11** 20 000

**12** **a** 5 **b** 13 **13** **a** 4 **b** 6 **c** 4

**14** square pyramid

## 6:4

**1** **a** 293 km **b** 330 km **2** **a** yes **b** yes **3** 1470 **4** 1502

**5** 370 **6** 13 and 16.

### Challenge

Answers will vary.

### Activity

**1** 4, 16, 100, 36, 81, and 49. **2** **a** 4, 16, 100, 36, and 14. **b** 45, 100, and 35. **c** 21, 35, 49, and 14.

## 7:1

**1** 27 **2** 28 **3** 25 **4** 42 **5** 10 052 **6** 28 **7** 3 **8** 50

**9** 1 **10** 20 652 **11** **a** 6 **b** 8 **12** **a** 16 **b** 9 **13** $\frac{3}{2}$

**14** **a** 72% **b** 9%

**15**

**16**

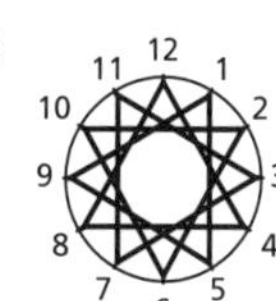



A 12-point star

## 7:2

**1** 7 **2** 99 **3** 5 **4** 40 **5** 5408 **6** 162 **7** 475 **8** \$79

**9** 29 **10** 8296 **11** **a** 60% **b** 30% **12** 96 408 **13** **a** 1 **b** 0

**14** **a** cylinder **b** cone **15** **a** 64 **b** 36

**16** $\frac{7}{5}$ $1\frac{2}{5}$ **17** **a**

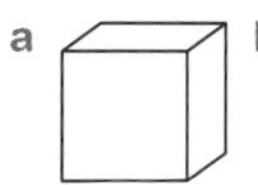

**b**

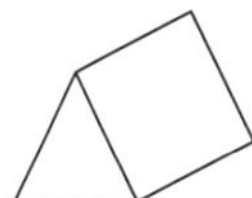

### Activity

**a** 7 **b** 8 and 3 **c** 14

## 7:3

**1** 312 **2** 68 **3** 65 **4** True **5** 92 485 **6** **a** 81 **b** 49

**7** **a** Students will copy the solid. **b** Students will copy the solid.

**8** $3\frac{3}{4}$

**9**

| Number of years | 1 | 2 | 3 | 4 | 5 |
|---|---|---|---|---|---|
| Number of months | 12 | 24 | 36 | 48 | 60 |

**a** *number of years* × 12 **b** 360

**10** **a**

| | | | | | | | | | |
|---|---|---|---|---|---|---|---|---|---|
| | | | | | | | | | |
| | | | | | | | | | |

**b** $\frac{14}{10}$

**11** $\frac{5}{4}$ **12** **a** 56% **b** $\frac{56}{100}$

## 7:4

**1** 49, 16, and 49 − 13 will be circled. **2** **a** $\frac{1}{10}$ **b** 0·1 **c** 10% **d** 90% **3** D **4** **a** 20 cm **b** 60 cm **5** D

### Challenge

Answers will vary.

### Activity

**a** 10 o'clock **b** no **c** no **d** no

## 8:1

**1** 22 **2** 48 **3** 39 **4** 71 **5** 5954 **6** 8 **7** 37 **8** $88

**9** 9 **10** 9280 **11** 45% **12** **a** 25 **b** 36 **13** rectangular pyramid

**14** **a** 100 **b** 49 **15** $6\frac{1}{3}$ **16** −17°C **17** **a** 10 **b** Rachel **c** Heather

## 8:2

**1** 80 **2** 27 **3** 27 **4** $\frac{17}{10}$ or $1\frac{7}{10}$ **5** 10456 **6** −7, −6 **7** 134 **8** 44

**9** $90 **10** 21632 **11** 78318 **12** 1, 4, 9, 16, 25, 36, 49, 64, 81, 100

**13** **a** 83% **b** 0·83 **14** **a** 9 **b** 64 **15** $\frac{7}{4}$

**16** −19°C **17** 1, 56, 2, 28, 4, 14, 7, 8 **18** 12

**19**

| Squares | 1 | 2 | 3 | 4 | 5 |
|---|---|---|---|---|---|
| Sides | 4 | 8 | **12** | **16** | **20** |

**20** Number line: [−4] −3 [−2] [−1] 0

### Activity

**a** 25 **b** 9 **c** 100 **d** 16 **e** 64 **f** 1 **g** 81 **h** 4 **i** 36

## 8:3

**1** 149 **2** 175 r 2 **3** 118 **4** $3\frac{4}{5}$ **5** 27671

**6** 68, Rule: 1st number + 15 **7** −8, −2, 0, 2, 4 **8** 25, 36, 49, 64

**9** **a** 70% **b** 10% **c** 7% **d** 1% **10** 807, <u>813</u>, 819, 825, <u>831</u>, 837

**11** 17 **12** Number line: −0·8 −0·6 −0·4 [−0·2] [0]

**13** −23°C **14** 1, 81, 3, 27 and 9. **15** **a** 0·38 **b** 38% **16** 242

## 8:4

**1** 25 **2** **a** 12 **b** 15 **3** **a** 21 cm **b** 84 cm **4** **a** 14 **b** 36

**5**

| □ | 15 | 25 | 35 | 44 | 54 |
|---|---|---|---|---|---|
| △ | 54 | 64 | 74 | 83 | 93 |

### Challenge

Answers may vary. See ID card D, Cards 4–14 for examples.

### Activity

**a** 21 **b** 2 **c** 0 **d** 18 **e** 11 **f** 3 **g** 25 **h** 7 **i** 48 **j** 2 **k** 24

## 9:1

**1** 77 **2** 132 **3** 60 **4** 43 **5** 2926 **6** 10 **7** 69 **8** $9

**9** 5 **10** $26840

**11** Number line: [−20%] [−15%] −10% −5% 0 **12** −12°C

**13** **a** $105 **b** $465 **14** **a** B **b** A **15** −7 **16** **a** −7 **b** −4 **c** −10

## 9:2

**1** 9 **2** $9 **3** 76 **4** $243 **5** $8373 **6** 165 **7** 52 **8** 356

**9** −2 **10** $19428 **11** 63929 **12** −49°C **13** **a** 10 **b** Monday and Thursday **14** obtuse angle **15** −9, −7, −2, 4, 8

**16** $4\frac{4}{5}$ **17** $0·29 = 29\% = \frac{29}{100}$ **18** 53

### Activity

(16) parallel lines (17) perpendicular lines (18) corner (19) arm (20) acute angle (21) right angle (22) obtuse angle (23) straight angle (24) reflex angle (25) revolution

## 9:3

**1** 121 **2** 324 r 1 **3** 46 **4** 3

**5** **a** 4 > −5 **b** −8 < 3 **c** −3 > −21 **d** −7 < −1

**6**

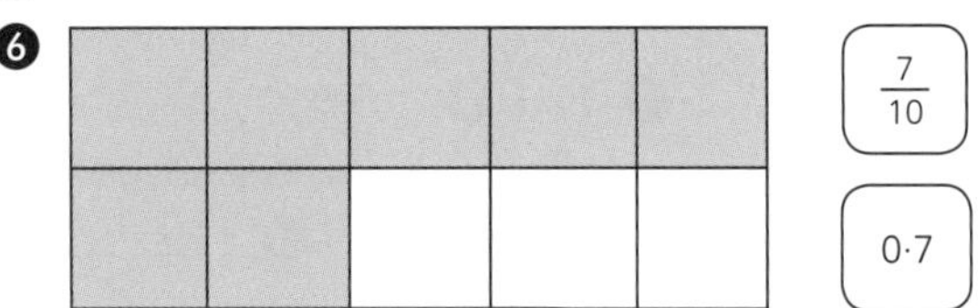


**7** **a** $\frac{20}{100}$ or $\frac{1}{5}$ **b** 27% **c** 120 **8** **a** 0·51 **b** 51% **9** **a** 3 squares will be coloured red. **b** 4 squares will be coloured blue. **c** no

## 9:4

**1** 4020 min **2** 8 **3** **a** C **b** B **c** D **d** E **4** 74 **5** 70° **6** 11

**7** 8 **8** Yes **9** 76

### Challenge

Answers will vary. E.g. 40 + 16, $7^2$ + 7, etc.

### Activity

23, 14, 47, 29·5, 35·5, 29, 18, 21, 27, 36

18, 15, 23·5, 24, 17, 19, 13, 12, 14, 16

□ = 7

## 10:1

**1** 50 **2** 11 **3** 7 **4** 25 **5** 7939 **6** 600 **7** 250 **8** 158

**9** 8 **10** 24188 **11** **a** 150° **b** obtuse

**12** 56350356, 56480093, 56798453

**13** $\frac{3}{5}$

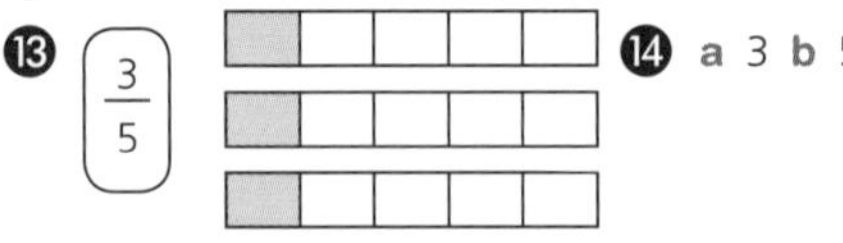

**14** **a** 3 **b** 5

**15** **a** −2 **b** 1 **c** −5 **16** 503054

## 10:2

**1** 902 **2** 680 **3** 12 **4** 6 **5** 50178 **6** $3 **7** 281 **8** 27

**9** 22 **10** $42399 **11** 180° **12** **a** −40% **b** 0% **c** −100%

**13** **a** About 22¥ **b** About 52¥ **c** About 46c **d** About 69c

**14** **a**

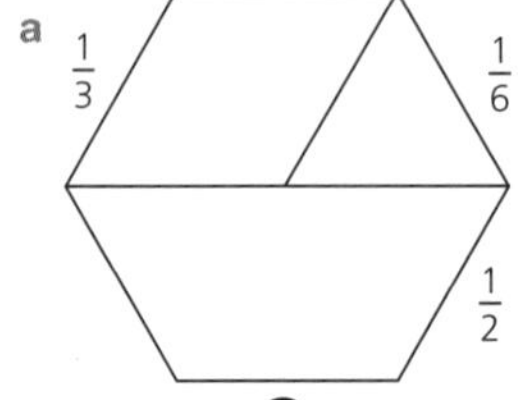


**b** yes **c** no **15** **a** 3 **b** 4



 • *AUSTRALIAN SIGNPOST MATHS 6 MENTALS* • ISBN 978 0 6557 0886 5

**Activity**

(1) integers (2) prime number (3) composite number (4) 7 (5) 2 and 4 (7) remainder (8) square numbers (9) digits (10) 4 ones, 3 tenths, 9 hundredths, 5 thousandths

## 10:3

**1** 124 **2** 121 **3** 94 **4** 34 067 **5** $3\frac{1}{2}$

**6** **a** $-3 > -7$ **b** $6 > -5$ **c** $0 > -1$ **d** $-10 < -8$

**7** **a** 0 **b** 1 **c** $-1\frac{1}{2}$ **8** **a** 12 **b** 20 **c** 18 **d** 16

**9** **a** G **b** L **c** L **d** H **10** 67 864

## 10:4

**1**

**2** 26 **3** **a** 7 **b** 3 **4** **a** 260 **b** 10 **5** **a** 9 **b** 12 **c** 150

**Challenge**

Answers may vary. E.g. This is an acute angle. Its size is between 0° and 90°. An open pair of scissors might have an acute angle.

**Activity**

41, 15, 18, 30, 5, 39, 14, 36, 23, 57

15, 42, 55, 43, 21, 34, 24, 30, 9, 29

even − odd = odd

odd − odd = even

## 11:1

**1** 240 **2** 270 **3** 132 **4** 6 **5** 4921 **6** 2400 **7** 1500

**8** 72 **9** $\frac{7}{9}$ **10** 4212 **11** $\frac{6}{10}$, 0·6 **12** $1800

**13** **a** 6 **b** 12 **c** 3 **d** 15 **14** **a** the flower **b** the tree **15** 1111

**16** 48 265 980, 48 376 098, 48 576 386

## 11:2

**1** 1400 **2** 70 **3** 90 **4** 919 **5** 7343 **6** 64 **7** $\frac{2}{10}$ or $\frac{1}{5}$

**8** 88 **9** 115 **10** 30 952 **11** **a** 6 **b** 3 **c** 15 **d** yes **12** $17.70

**13** **a** the person in the aisle **b** the exit **c** the spider **d** the movie screen **e** the audience

**Activity**

6, 21, 12, 18, 3, 30, 27, 9, 24, 15

12, 42, 24, 36, 6, 60, 54, 18, 48, 30

## 11:3

**1** 194 **2** 78 **3** 34 **4** 1652 **5** Approximately: **a** 207 km **b** 306 km **c** 351 km **d** 252 km **6** 87 **7** **a** 4 **b** 12 **c** 2 **d** 14

**8** **a** right angle **b** trapezium **c** reflection **d** pentagon **e** translation **f** rotation **g** parallel lines **h** parallelogram **9** 9 teams of six, with 5 children remaining **10** 800

## 11:4

**1** **a** $2.75 **b** $71.50 **2** 5 weeks **3** 800 m **4** **a** 20 **b** $16\frac{1}{2}$

**5** **6** 9

**Challenge**

Answers may vary. E.g. The C is west of the star.

**Activity**

**a** 1500 **b** 5600 **c** 2400 **d** 2500 **e** 4500 **f** 5000 **g** 4000 **h** 12 000 **i** 2500 **j** 640 000

## 12:1

**1** 93 **2** 1 **3** 48 **4** 56 **5** 8644 **6** 60 **7** 10 **8** 81

**9** 53 **10** 5120 **11** 768 L **12** 676 000 000

**13** **a** 82 576 354 < 82 586 739 **b** 98 735 089 < 98 735 900

**14** **a** kite **b** perpendicular lines **15** **a** 108 **b** 77 **c** 6000

**16** **a** 6000 **b** 14 000 **c** 970 **d** 18 000 **e** 670 **f** 6000

## 12:2

**1** 536 **2** 440 **3** 536 **4** $\frac{7}{6}$ or $1\frac{1}{6}$ **5** 5622 **6** 700 **7** 63

**8** 1134 **9** 439 **10** 14 686 **11** **a** Kyrgyzstan **b** Afghanistan **c** E3 **d** B3 **e** 800 km **f** Bishkek **g** Amritsar **12** 46

**Activity**

2, 6, 4, 8, 10, 5, 9, 3, 7, 1

3, 8, 5, 2, 7, 4, 1, 9, 6, 10

## 12:3

**1** 99 **2** 32 **3** 43 **4** 1012 **5** **a** i **b** n **6** 2·688 L

**7** **a** 5 **b** 10 **c** 3 **d** 12 **8** **a** kite **b** 3D **9** 12·7 km

## 12:4

**1** 316 m **2** 144 **3** 6 **4** **a** 2000 **b** 6000 **c** 10 000

**5** **a** Jiyu **b** 1601 m **6** **a** 4 **b** 327

**Challenge**

Answers will vary.

E.g. $6 \times 8 \div 2$, $4^2 + 2^2 + (9 - 5)$, etc.

**Activity**

**a** 100 cm **b** 175 cm **c** 130 cm **d** 190 cm **e** 60 cm

(Answers may vary. Here we measure from the base of the heel to the top of the head (not including hat or dog's ears, etc.).)

## 13:1

**1** 2400 **2** 4000 **3** $\frac{5}{10}$ or $\frac{1}{2}$ **4** 618 **5** 153 180 **6** 214

**7** 56 **8** 266 **9** 96 **10** 382 496 **11** 335 **12** **a** about 20 m **b** east **c** south-westerly **d** Year 1 **13** itself

## 13:2

**1** 200 **2** 60 **3** $3\frac{1}{4}$ **4** 9 **5** 46 391 **6** 133 **7** $\frac{5}{8}$ **8** 180

**9** 112 **10** 563 346 **11** 21 **12** **a** west **b** north-east **13** 112 000

**14** **a** 310° **b** 135° **15** **a** 700 **b** 17

**Activity**

1, 8, 5, 2, 7, 9, 3, 10, 6, 4

5, 1, 10, 6, 2, 8, 4, 7, 3, 9

□ = 8

□ = 8

## 13:3

❶ 73 r 2 ❷ 189 r 2 ❸ 52 r 8 ❹ **a** 15·4 L **b** 2600 mL
❺ **a** 1000 km **b** 3300 km **c** ACT, Tas, Vic, NSW, SA, NT, Qld, WA
❻ 2·5 km ❼ 142° ❽ 12 ❾ $1142.85 ❿ 793·06
⓫ 54 000 000

## 13:4

❶ **a** 842 500 **b** 325 000 ❷ **a** $34.80 **b** $2.90 ❸ 8
❹ 156 BCE ❺ 1 ❻ around 1000 times if you sleep for 8 hours
❼ 240

### Challenge

Answers will vary.
E.g. $10^2$, $58 \times 2 - 4^2$, etc.

### Activity

12, 42, 18, 48, 60, 36, 24, 54, 30, 0
18, 63, 27, 72, 90, 54, 36, 81, 45, 9

## 14:1

❶ 257 ❷ $\frac{5}{6}$ ❸ 422 ❹ 5 ❺ 8285 ❻ 32 ❼ 180 ❽ 41
❾ 73 ❿ 15 000 ⓫ **a** 12 r 1 **b** 8 r 6 ⓬ **a** 140° **b** 125°
⓭ **a** North **b** 90° ⓮ $1643.50 ⓯ 367 ⓰ −8°C ⓱ 600 000

## 14:2

❶ 317 ❷ 288 ❸ $6\frac{2}{3}$ ❹ 81 ❺ 11 324 ❻ 648 ❼ $\frac{9}{6}$ or $1\frac{1}{2}$
❽ 1143 ❾ 351 ❿ 685 020 ⓫ **a** 5 r 2 **b** 6 r 3 ⓬ 812 mL
⓭ 4 004 047 ⓮ 145° ⓯ $17 845
⓰ **a** 6 after 4 **b** 28 to 10
⓱ **a** $1\frac{2}{3}$ **b** $\frac{5}{3}$ ⓲ 1000

### Activity

(15) regular shapes (16) irregular shapes (17) diagonals (18) axis of symmetry (19) axes of symmetry (20) rotational symmetry (21) reflection (22) translation (23) rotation (26) tessellation

## 14:3

❶ 72 r 3 ❷ 100 r 4 ❸ 37 r 6 ❹ 132 r 3 ❺ 71 r 1 ❻ 101 r 1
❼ 90 ❽ **a** 75° **b** 100° ❾ 4 ❿ 0·004 ⓫ **a** 4 **b** 2 ⓬ 247 368
⓭ 1 097 365, 1 138 000, 1 277 456 ⓮ **a** 6800 g **b** 6·7 cm

## 14:4

❶ 101 CE ❷ **a** 45° **b** 135° ❸ **a** $85 410 **b** $124 554
❹ (18 − 5) × 3 = 39 ❺ 624

### Challenge

Answers may vary. E.g. It is an eight-digit even number. It becomes 79 000 000 when rounded to the nearest million.

### Activity

(5) rectangle (6) rhombus (7) parallelogram (8) trapezium (9) kite (10) quadrilaterals (11) pentagons (12) hexagons (13) octagon (14) decagon

## 15:1

❶ 96 ❷ 2800 ❸ 30 ❹ 14 ❺ $9451 ❻ $4.57 ❼ $4.80
❽ $0.60 ❾ 1 ❿ $5435 ⓫ 45° ⓬ $1\frac{1}{4}$
⓭ **a** Our team total = 26, Our opposition total = 14 **b** $4\frac{2}{6}$ or $4\frac{1}{3}$
**c** $2\frac{2}{6}$ or $2\frac{1}{3}$ **d** 4 ⓮ **a** J **b** P **c** (1, 2) **d** (1, 1)

## 15:2

❶ 510 ❷ 78 ❸ 0 ❹ $476 ❺ 9470 ❻ 67 ❼ 163 ❽ 137
❾ 72 ❿ 29 192 ⓫ 72 cm ⓬ 60° ⓭ 35·025 cm ⓮ 6
⓯ **a** 90° **b** 135° ⓰ **a** A **b** E **c** (2, 2) **d** (3, 5)

### Activity

**a** 9 m **b** 97 g **c** 606 **d** 6 **e** 63

## 15:3

❶ **a** $8\frac{3}{5}$ **b** $4\frac{2}{8}$ or $4\frac{1}{4}$ **c** $139\frac{1}{6}$ **d** $195\frac{2}{4}$ or $195\frac{1}{2}$ **e** $60\frac{6}{9}$ or $60\frac{2}{3}$
**f** $73\frac{6}{10}$ or $73\frac{3}{5}$ ❷ revolution
❸ $1\frac{2}{6}$ or $1\frac{1}{3}$ ❹ **a** 320° **b** 240° ❺ 240 km ❻ **a** B **b** I
**c** (2, 2) **d** (−2, −4) **e** (−3, −3) **f** (−3, 4) ❼ 18 995
❽ **a** $\frac{9}{4}$ **b** $\frac{18}{5}$ ❾ 0·005 or 5 thousandths

## 15:4

❶ **a** $6\frac{1}{3}$ **b** 7 **c** $7\frac{1}{2}$ **d** $8\frac{1}{3}$ ❷ 8185 mL or 8·185 L ❸ $1169
❹ $128 047.62 ❺ 32 ❻ 2·658 $m^2$

### Challenge

Answers will vary.
E.g. Group: $13, $84, $56, $17, $35 Average: $41

### Activity

14, 42, 7, 28, 70, 49, 21, 63, 35, 56
16, 48, 8, 32, 80, 56, 24, 72, 40, 64

## 16:1

❶ 560 ❷ 6 ❸ 13 ❹ 60 ❺ 78 216 ❻ 87 ❼ 43 ❽ 343
❾ $\frac{7}{2}$ or $3\frac{1}{2}$ ❿ 49 060 ⓫ $1\frac{2}{3}$ or $1\frac{4}{6}$ cucumbers ⓬ 120 km
⓭ **a** and **b**

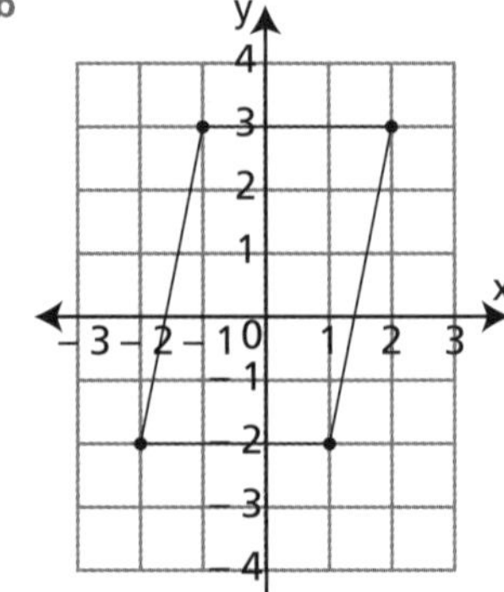

**c** parallelogram

⓮ **a** triangular prism **b** triangle **c** H and E **d** A, C, G, D and B.

## 16:2

❶ 920 ❷ 309 ❸ 16 ❹ 0 ❺ 456525 ❻ $\frac{6}{12}$ or $\frac{1}{2}$
❼ 1292 ❽ 3 ❾ 40 ❿ 378949 ⓫ 23 km ⓬ 92746
⓭

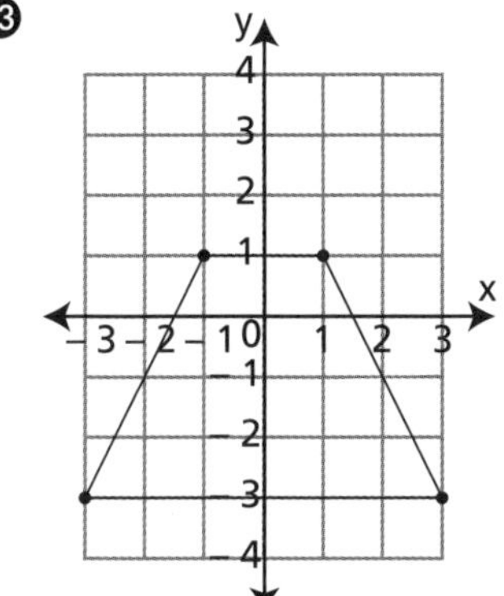

c trapezium
⓮ a 178110 b yes ⓯ 201 g
⓰ Cube and square pyramid, square prism.

**Activity**
1, 2, 3, 4, 5, 6, 7, 8, 9, 10
5, 8, 1, 10, 4, 6, 2, 9, 3, 7

## 16:3

❶ a $8\frac{1}{7}$ b $18\frac{4}{5}$ c $231\frac{2}{3}$ d $115\frac{1}{2}$ e $79\frac{7}{8}$ f $39\frac{2}{10}$ ❷ 73868
❸ 91153 ❹ 2 km ❺ a 120° b 180° ❻ $3\frac{2}{3}$ slices ❼ 114·75 g
❽ $366550 ❾ 57 ❿ A cube has 6 faces and 12 edges.

## 16:4

❶ a 2 b 3 ❷ 480 km ❸ 15 ❹ 130 ❺ a 9044 b 17550
❻ 90

**Challenge**
Answers will vary.
E.g. Rule = × 16 + 17

| 1st number | 1 | 2 | 3 | 4 | 5 | 6 | 7 |
|---|---|---|---|---|---|---|---|
| 2nd number | 33 | 49 | 65 | 81 | 97 | 113 | 129 |

**Activity**
a $12.50
b Loss of $7400
c Profit of $6870

## 17:1

❶ 562 ❷ 48 ❸ 9 ❹ 9 ❺ 535502 ❻ 181 ❼ 60 ❽ 630
❾ 25 ❿ 780670 ⓫ $1552 ⓬ This is the net of a square pyramid.
⓭ a square ⓮ a $\frac{1}{10}$ b 0·1 c 10% ⓯ a no b Jessica
⓰ 1·4, 1·6, 1·8, 2·0 ⓱ $\frac{19}{5}$ ⓲ a 5 b 12 ⓳ 0·006 or $\frac{6}{1000}$

## 17:2

❶ 5162 ❷ 921 ❸ 644 ❹ 80 ❺ 786352 ❻ 9 ❼ $31
❽ $\frac{8}{10}$ or $\frac{4}{5}$ ❾ 65% ❿ 873522 ⓫ 85% ⓬ 840 km
⓭ 10127 km ⓮ a Students will copy the shape. b 9 vertices
c 9 faces d squares ⓯ This is the net of a cone.
⓰ a triangle ⓱ 0·09 or $\frac{9}{100}$

**Activity**
Answers will vary. E.g. 2·5 kg of Apples + 1 kg of Pears + 1 Kiwi Fruit + 3 kg of Oranges + 0·5 kg of Bananas

## 17:3

❶ 4213213 ❷ 864025 ❸ 2356 m ❹ This is the net of a cube.
❺ a hexagon ❻ a $\frac{1}{2}$ b 0·5 c 50% ❼ a 4% b $\frac{4}{100}$ or $\frac{1}{25}$ c 0·04
❽ a 2·0, 2·6, 3·2, 3·8 b 1·2, 1·5, 1·8, 2·1 ❾ 0·9 or $\frac{9}{10}$ ❿ 5400
⓫ a B, E. b B

## 17:4

❶ a 30 b 12 ❷ 38 ❸ 25 ❹ a true b true
❺ a 81 b 162 c 216 d 206 ❻ 9174·732

**Challenge**
Answers will vary.
a 988308 + 786000, 1024308 + 750000, etc.
b 980000 + 243107, 1173107 + 50000, etc.

**Activity**
A about 0·5 B Estimates will vary. C Estimates will vary.
D about 0·17 E 0

## 18:1

❶ 6 ❷ 7 ❸ 2 ❹ 70 ❺ 391166 ❻ 80 ❼ 6 ❽ 21
❾ $651 ❿ 762487 ⓫ a Blue b Green c Green d Blue ⓬ 27
⓭ $731\frac{2}{5}$ mL or 731·4 mL ⓮ Total spent = $5.90,
Amount left = $24.10 ⓯ a $\frac{5}{6}$ b $\frac{5}{12}$

## 18:2

❶ $91\frac{1}{9}$ ❷ 5 ❸ 80 ❹ 60 ❺ 310023 ❻ 462 ❼ 390
❽ $644.20 ❾ $50.45 ❿ $12.10 ⓫ A head and a tail.
⓬ B ⓭ a $182 b $180 ⓮ $25.05 ⓯ 7

**Activity**
2, 6, 3, 9, 7, 4, 10, 8, 5, 1
2, 5, 8, 1, 7, 9, 3, 6, 4, 10
How many weeks? 4

## 18:3

❶ 13841429 ❷ $324.60 ❸ a G b B or Y ❹ 280 km
❺ $14.30 ❻ a 18 b 41 c 59 d 0 e 1 f 6 g 11 h $\frac{1}{2}$ i $\frac{5}{2}$ or $2\frac{1}{2}$
❼ There is a skew towards the lower scores.

## 18:4

❶ 16 ❷ a 19 cm b 57 cm ❸ 1980 km ❹ 769·5 km
❺ a 156 b 386 c 116 d 277

**Challenge**
Answers will vary.
E.g. There is a 50% chance of tossing a head on a coin.

**Activity**
Answers will vary.
E.g. 1500 g of Crumpets + 1200 g of Honey + 2000 g of Jam + 510 g of Cheese Spread

## 19:1

❶ 210 ❷ 3200 ❸ 32 ❹ 6 ❺ 0·85 ❻ 5 ❼ 0·7 ❽ 0·9
❾ 23 ❿ 6·60 ⓫ **a** $6 will be circled. **b** $10.80 **c** $15 will be circled.
⓬ 2, 2, 2, 3, 4, 5, 6, 8 **a** 6 **b** 3·5 **c** 2 **d** 4 ⓭ 7·21, 8·9, 54, 67·25

## 19:2

❶ $456 ❷ $1560 ❸ 496 ❹ 40 ❺ 101·519 ❻ 20 ❼ 8
❽ 32 ❾ 26 ❿ 2·2 billion ⓫ $163.10 ⓬ 336 km
⓭ **a** 13 **b** 6 **c** 7 **d** $9\frac{1}{7}$ ⓮ **a** 1 **b** 2 ⓯ **a** Wiping benches **b** 35

**Activity**

Alan's age = 51

## 19:3

❶ 13·037 ❷ 11·443 km ❸ Total amount spent = $52.20, Amount left = $17.80 ❹ **a** 12 **b** 40 **c** 38 **d** 38·7
❺ $10.2 million
❻

| Cups of water we drank | | Number |
|---|---|---|
| Jasmine | 𝍸 IIII | 9 |
| Matilda | 𝍸 𝍸 II | 12 |
| Lydia | 𝍸 II | 7 |
| Isaac | 𝍸 𝍸 𝍸 I | 16 |

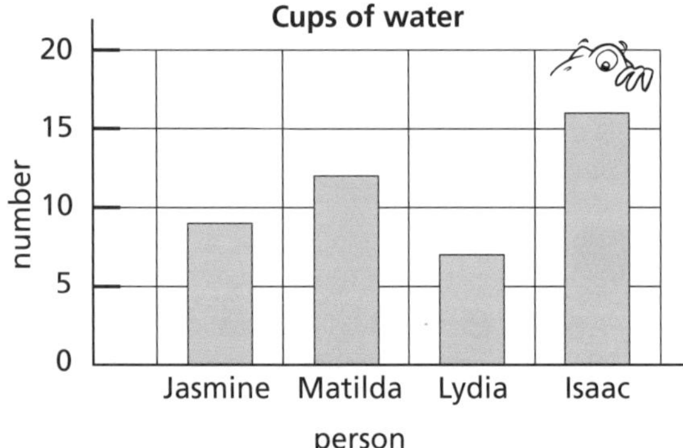


## 19:4

❶
```
    7 ·[3] 7  5
+  [9]· 8  0  6
---------------
 1  7 · 1 [8] 1
```
❷
```
    2 · 9  5  7
+  [7]· 4 [ ] 6
---------------
 1  0 · 4 [9]5 3
```
❸ 48

❹ **a** New Zealand **b** 10 medals **c** 0·57 **d** 3·31 ❺ 265 BCE

**Challenge**

Answers may vary. E.g. The title implies that all Australians were represented in the survey.

**Activity**

**a**

| 1 | | | |
|---|---|---|---|
| 2 | 3 | 4 | 5 |
| 6 | 7 | 8 | 9 |
| 10 | 11 | 12 | 13 |
| 14 | 15 | 16 | 17 |
| 18 | 19 | 20 | 21 |
| 22 | 23 | 24 | 25 |
| 26 | 27 | 28 | 29 |
| 30 | 31 | 32 | 33 |

**b** Which numbers are left? 1, 2, 4, 8, 16, 32.
What kind are they? Powers of 2.

## 20:1

❶ 5 ❷ 10 ❸ 9 ❹ 5 ❺ 493·09 ❻ 44 ❼ 33 ❽ 2 ❾ 90
❿ 93·482 ⓫ $5.2 million ⓬ $5 ⓭ **a** 8000 m **b** 7·655 L
**c** 4·5 cm **d** 3·56 m ⓮ **a** 12·4 m **b** 2·8 m ⓯ 32·23 kg
⓰ 995·5 g ⓱ $5\frac{1}{4}$, 4, $3\frac{1}{8}$, $1\frac{3}{8}$

## 20:2

❶ 373 ❷ 549 ❸ 21 ❹ 25 ❺ 192·496 ❻ 20
❼ $18 ❽ 8500 cm ❾ 1234 ❿ 88·113 ⓫ 71
⓬ 201 ⓭ 81 r 5 ⓮ $176.25 ⓯ 62·302 L
⓰ Answers may vary. E.g. The title doesn't show that only the cricket team were surveyed.
⓱ 998·6 m ⓲ They tessellate. ⓳ 12 m

**Activity**

66, 44, 59, 38, 52, 47, 31, 25, 20, 42
70, 56, 52, 55, 48, 51, 27, 29, 51, 73

## 20:3

❶ 9·875 ❷ 0·1195 km ❸ **a** 10·214 km **b** 0·48 km
❹ **a** B **b** 0·5 **c** 0·25 ❺ 3·88 kg ❻ **a** $\frac{1}{2}$ **b** $\frac{1}{6}$ **c** $\frac{2}{6}$ or $\frac{1}{3}$ **d** $\frac{3}{4}$
❼ A was reflected about its right side.
❽ 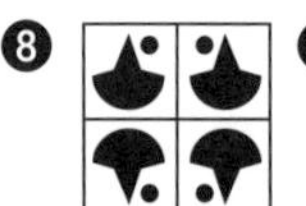 ❾ 34 927

## 20:4

❶
```
    8 ·[2] 3  9
+  [3]· 9 [7] 3
---------------
 1  2 · 2  1  2
```
❷
```
    2 · 9  5 [9]
+  [9]· 4 [6] 6
---------------
 1  2 · 4  2  5
```

❸ 52 ❹ 4995·455 kg ❺ 30 ❻ 310 ❼ 0·545 kg or 545 g

**Challenge**

Answers will vary. One example is:

**Activity**

65, 29, 56, 37, 68, 34, 53, 32, 71, 50
37, 36, 39, 68, 50, 57, 45, 54, 62, 81
even + odd = odd
odd + even = odd

## 21:1

❶ 1 ❷ $350 ❸ 3 ❹ 4 ❺ 16·117 ❻ 97·5 m ❼ 6400
❽ 0 ❾ 12 ❿ 45·80 ⓫ 98·9 cm ⓬ **a** $\frac{3}{4}$ **b** $1\frac{1}{8}$ **c** $1\frac{1}{2}$ **d** $2\frac{3}{8}$

**13** a 9, 9·343 b 8, 7·893 **14** a $56.90 b $57 **15** 9·9

**16** a 10 b 20 **17** 6546 g

## 21:2

**1** 90 **2** 6 **3** 70 **4** $\frac{3}{10}$ **5** $903.23 **6** $3920 **7** 69

**8** 1935 **9** 56 **10** $33 474 **11** 56 **12** 64 **13** 64

**14** Shape A was translated to the right. **15** 15·35 m

**16** a 14, 14·065 b 12, 12·064 **17** 45·33

**18** a 5600 b 6900 c 8·8

### Activity

(5) rectangle (6) rhombus (7) parallelogram (8) trapezium (9) kite (10) quadrilaterals (11) pentagons (12) hexagons (13) octagon (14) decagon

## 21:3

**1** 6, 6·278 **2** 13, 12·329 **3** −3, −2, 5 **4** 5 **5** 0·06, 0·36, 0·63

**6** 108 **7** 67 **8** 298·106 **9** a south-west b north-east

**10** 1·85 **11** 33 **12** −12°C **13** a $\frac{75}{100}$ or $\frac{3}{4}$ b 0·75

## 21:4

**1**

**2**

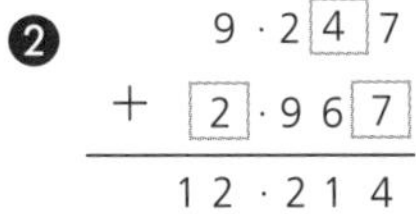


**3** a 15 b 30 **4** 325 BCE **5** 5 **6** 36

**7** a 355·6 km b 2133·6 km

### Challenge

Answers will vary. See ID Card C, tile 26 for an example.

### Activity

| | | | | | | |
|---|---|---|---|---|---|---|
| 1 2 | 4 | 2 7 | 0 | 3 5 | ■ | 4 8 |
| 3 | ■ | 3 | ■ | 5 5 | 4 | 0 |
| 6 4 | 7 3 | 6 | 8 | ■ | ■ | 4 |
| ■ | 8 1 | 9 | ■ | 9 6 | 6 | 9 |
| ■ | 3 | ■ | ■ | 0 | ■ | 0 |
| ■ | 10 6 | 7 | 4 | 0 | 4 | 1 |

**Across:**

**1.** 24 705 **5.** 540 **6.** 4368 **8.** 19 **9.** 669 **10.** 674 041

**Down:**

**1.** 234 **2.** 7369 **3.** 55 **4.** 804 901 **7.** 3136 **9.** 600

## 22:1

**1** 11·8 **2** 132 **3** 30 **4** 16 **5** $74.34 **6** 6300 **7** 24

**8** 70 **9** $\frac{7}{10}$ **10** $137.70

**11** a 2 (counting numbers) b 5 (counting numbers)

**12** a 3·6 b $3\frac{6}{10}$ **13** 4·2 L **14** 2, 3, 4, 5, 6, 7, 8, 9, 10, 11, and 12.

**15** a $\frac{9}{4}$ b $\frac{18}{5}$ **16** 75 921 977, 75 241 119, 75 124 911

**17** 62·45, 64·48

## 22:2

**1** 594 **2** 50 **3** 571 **4** 35 **5** $101.46 **6** 6700 km **7** 1

**8** $\frac{17}{12}$ or $1\frac{5}{12}$ **9** 891 **10** 319·14 **11** 162 **12** 133 r 5 **13** 96 r 5

**14** 6·4 m **15** 50% **16** 4108 paces. **17** 43·6 m

**18** 1, 18, 2, 9, 3 and 6. **19** a 10, 10·091 b 11, 10·986 **20** $450

### Activity

a

| | | |
|---|---|---|
| 8 | 3 | 4 |
| 1 | 5 | 9 |
| 6 | 7 | 2 |

b

| | | |
|---|---|---|
| 9 | 4 | 5 |
| 2 | 6 | 10 |
| 7 | 8 | 3 |

c

| | | |
|---|---|---|
| 8 | 3 | 10 |
| 9 | 7 | 5 |
| 4 | 11 | 6 |

## 22:3

**1** a 10, 10·005 b 10, 9·673 **2** 407 cm or 4·07 m

**3** a 0·02 or $\frac{2}{100}$ b 0·2 or $\frac{2}{10}$ **4** 330° **5** 4·2 L

**6** a $61.70 b $33.80 c $31.15 **7** 38 kg **8** 36

## 22:4

**1**

[8] · 3 [5] 9
× [7]
58 · 5 1 3

**2**

[9] · [5] 6 [8]
× 5
4 7 · 8 4 0

**3** 120 h **4** 14 **5** 20 **6** $49 **7** Estimates will be around 8967.

### Challenge

Answers will vary.

E.g.

a 3 × 3, (6 × 3) − $3^2$, etc.

b 6 × 10 + 4, $8^2$, etc.

### Activity

| | | | | | | |
|---|---|---|---|---|---|---|
| 1 2 | 3 | 2 5 | 6 | 3 2 | ■ | 4 5 |
| 4 | ■ | 1 | ■ | 5 5 | 4 | 6 |
| 6 4 | 7 5 | 9 | 3 | ■ | ■ | 9 |
| ■ | 8 1 | 3 | ■ | 9 8 | 6 | 5 |
| ■ | 8 | ■ | ■ | 0 | ■ | 4 |
| ■ | 10 4 | 5 | 5 | 6 | 2 | 5 |

**Across:**

**1.** 23 562 **5.** 546 **6.** 4593 **8.** 13 **9.** 865 **10.** 455 625

**Down:**

**1.** 244 **2.** 5193 **3.** 25 **4.** 569 545 **7.** 5184 **9.** 806

## 23:1

**1** 1 **2** 4·5 **3** $\frac{2}{4}$ or $\frac{1}{2}$ **4** 37 **5** 5·534 **6** 620 **7** 7

**8** 522 **9** 81 **10** 8·096 km **11** 430 r 5 **12** 6651 r 2

**13** 7·17 **14** 37 **15** a $\frac{19}{5}$ b $\frac{17}{8}$ **16** 7 000 000 **17** a 6500 b 7535

**18** a false b true **19** 9 **20** 56 cm **21** 83 **22** 0·06 or $\frac{6}{100}$

## 23:2

**1** $\frac{12}{10}$ or $1\frac{2}{10}$ **2** 18 **3** 3819 **4** 3366 **5** 43·848 m **6** 0·25
**7** 255 **8** 290 **9** $120 **10** 42·534 cm **11** 9689 **12** 1161
**13** $\frac{3}{6}$ or $\frac{1}{2}$ **14** 2 **15** Monday **16** 56 499 **17** 310° **18** 1501
**19** 345·6 cm **20** 5686

**Activity**

5, 7, 1, 12, 9, 17, 15, 21, 25, 27
1, 20, 32, 11, 3, 6, 4, 39, 28, 24
10, 14, 7, 32, 41, 24, 0, 19, 8, 30

## 23:3

**1** a 16, 15·963 b 14, 13·887 **2** 2546 **3** 3012 **4** $\frac{3}{6}$ or $\frac{1}{2}$
**5** 25 mL **6** 125 mL **7** 356·7 cm **8** 540 **9** $458
**10** a 467 000 m b 5·763 kg c 7800 kg d 4·657 m e 8·674 L
f 60 000 $m^2$ g 78 400 m **11** $43.65

## 23:4

**1** 1600 and 420 will be circled. **2** 59·8 **3** a $576 b $460.80
**4** 44 km **5** 504 times. **6** Thursday **7** 286 **8** 250

**Challenge**

Answers will vary.

**Activity**

| | | |
|---|---|---|
| 8·375 km | 8 km 375 m | 8375 m |
| 2·914 km | 2 km 914 m | 2914 m |
| 5·446 km | 5 km 446 m | 5446 m |
| 9·125 km | 9 km 125 m | 9125 m |
| 3·546 km | 3 km 546 m | 3546 m |
| 9·897 km | 9 km 897 m | 9897 m |

## 24:1

**1** 833 **2** 38 **3** 26 **4** 25 **5** 5·226 **6** $\frac{3}{10}$ **7** 376 **8** 8
**9** 44 **10** 21·072 **11** 745·9 **12** 9·03 **13** a 45·643 b 934·5
c 0·0706384 d 4656·78 **14** 3456 **15** 57 mm or 5·7 cm
**16** a 45 mm b 7·8 cm c 4·3 km d 6800 m **17** B

## 24:2

**1** 45 **2** 16 **3** 75 **4** 0·7896 **5** 24·472 m **6** 2 **7** $11.33
**8** $\frac{2}{10}$ or $\frac{1}{5}$ **9** 599 **10** 47·912 cm **11** 37 500·8 **12** 12·72
**13** a $4\frac{3}{10}$ b $\frac{43}{10}$ c 4·3 **14** a 2·1 cm b 21 mm **15** 22·5 km
**16** a 87·645 b 57 893·4 **17** a 500 m b 1750 m **18** 89 km/h

**Activity**

a Rachel b Rhonda c Heather

## 24:3

**1** 6 669 085 **2** 29·945 m **3** 265·8 **4** 2·99
**5** a 62·4721 b 46 567·8 **6** a 3·4 cm b 34 mm **7** 123 km/h
**8** 4·35 km or 4350 m **9** a 9700 m b 36·7 cm c 8·47 km
d 137 mm e 5400 m **10** 14
**11** 24 × 20 = (20 × 20) + (4 × 20) = 400 + 80 = 480

## 24:4

**1** 6000 **2** 30 min **3** 5 **4** 4·533 km or 4533 m **5** 9 **6** 8

**Challenge**

1st column: 6378·9, 4·9028, 8703·4, 941·6, 310·8, 9·0023, 832·5, 79·32
2nd column: 10·23, 324·1, 3·4098, 4·2937, 3000·2, 5112·3, 9·0076, 23·1

**Activity**

a 100

b

| Thickness | 10 mm | 1 mm | 0·1 mm |
|---|---|---|---|
| Number of sheets | 100 | 10 | 1 |

c 0·8 cm or 8 mm

## 25:1

**1** 21 **2** 741 **3** 2 **4** 26 **5** 269·98 **6** $\frac{8}{12}$ or $\frac{2}{3}$ **7** 367 km
**8** −4°C **9** 52 **10** 783·03
**11** a b

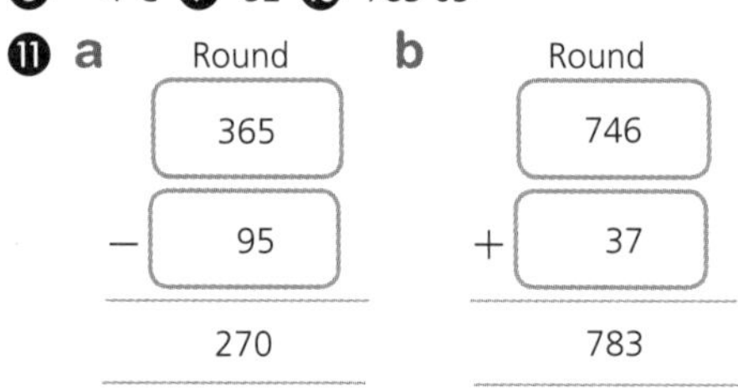


**12** a 50 ÷ 5 = 10 b 49 ÷ 7 = 7
**13** a Estimates will vary, between 40–50 mm
b Measure = 45 mm or 4·5 cm
**14** a 624·721 b 4656·78
**15** a 599 mm b 876·9 cm c 2500 m d 8200 m

## 25:2

**1** 54 **2** 20 L **3** 4·59 **4** $7\frac{1}{5}$ **5** 289 556 **6** $\frac{5}{12}$ **7** 4588
**8** 6 **9** $3.55 **10** 222 879 **11** 5·86 m
**12** a 90 ÷ 3 = 30 b 36 ÷ 6 = 6 **13** a 39 b 5 **14** a 3·566 b 948 230
**15** a Estimates will vary, between 70–80 mm
b Measure = 73 mm or 7·3 cm
**16** a 5690 cm b 87·654 km c 375·6 cm d 27 cm 9 mm e 8846 m

**Activity**

A 2 m B 8 cm C 10 m D 2 cm E 20 cm F 75 cm G 4 m H 0·5 cm

## 25:3

**1** 4 234 915 **2** $3141.65 **3** 9·26 m **4** 43·7 m
**5** a 64 ÷ 8 = 8 b 27 ÷ 9 = 3 **6** a 16 b 9 **7** $\frac{1}{12}, \frac{1}{6}, \frac{1}{4}, \frac{1}{3}, \frac{1}{2}$
**8** a 89·289 b 79 254 **9** a 3·527 b 8·721 c 9·138
**10** a 38 cm 2 mm b 36 670 m **11** 3810

## 25:4

**1** 15 564 **2** 330 mm **3** a 13 152 m b 13 min

### Challenge

Answers will vary.

### Activity

1, 12, 2, 6, 3, 4

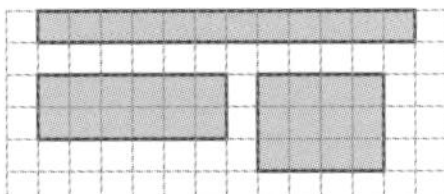

## 26:1

**1** 599 **2** 514 **3** 70 **4** $4\frac{1}{2}$ **5** 66 **6** 31 **7** 24 **8** $\frac{3}{12}$ or $\frac{1}{4}$
**9** 9·42 **10** 35 465·7 **11** a Estimates will vary, between 50–60 mm
b Measure = 57 mm or 5·7 cm **12** a 5 cm 6 mm b 78 km 720 m
**13** 4 **14** $\frac{7}{12}$ **15** a $\frac{3}{4}$ b $\frac{4}{4}$ **16** a 54·621 b 86 734
**17** a 57 b 8684 **18** 360

## 26:2

**1** 49 **2** 37 L **3** $\frac{1}{12}$ **4** \$38 **5** 51 **6** 52 **7** 84 **8** 42 mL
**9** 18·31 **10** 7108·9 **11** 3·75 m **12** 79·2 cm **13** 86·734
**14** 8 cm 3 mm **15** a Estimates will vary, between 20–30 mm
b Measure = 26 mm or 2·6 cm **16** 23 cm
**17** a 54 ÷ 9 = 6 b 56 ÷ 7 = 8 **18** a 30 b 8 **19** $\frac{6}{8}$ or $\frac{3}{4}$

### Activity

a 4 b $\frac{3}{4}$ c $\frac{5}{8}$ d $\frac{4}{8}$ or $\frac{1}{2}$ e $\frac{3}{8}$ f $\frac{9}{8}$ or $1\frac{1}{8}$ g $\frac{1}{8}$ h $1\frac{5}{8}$ i $1\frac{7}{8}$ j $\frac{7}{8}$

## 26:3

**1** 6288·32 **2** 209·88 **3** 161·2 cm or 1·612 m **4** a 80 b 7
**5** 0% **6** 167 **7** 3·677 **8** a $3\frac{1}{2}$ b $6\frac{1}{2}$ **9** 9
**10** a 62 973 b 83 497 **11** 12 **12** 410·3 **13** a $\frac{1}{5} < \frac{3}{10}$ b $\frac{7}{10} > \frac{2}{5}$
**14** 78 984

## 26:4

**1** 528 cm or 5·28 m **2** \$94.95 **3** \$3 086 800
**4** 1 and 14; 2 and 7 **5** 978
**6**

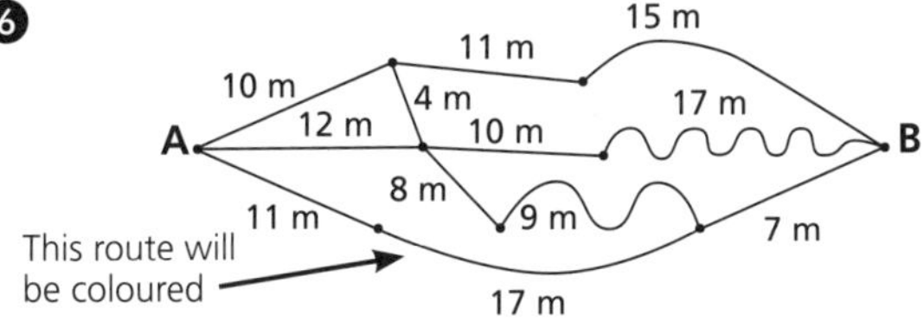


### Challenge

Answers may vary. E.g. It is a 5 digit number. It has 1 ten, 3 ones, 4 tenths, 5 hundredths, and 2 thousandths. It becomes 13 when rounded to the nearest whole number.

### Activity

a $3\frac{1}{2}$ b $1\frac{1}{2}$ c $2\frac{1}{4}$ d $2\frac{3}{4}$ e $\frac{3}{4}$ f $\frac{1}{4}$ g $3\frac{1}{2}$ h $1\frac{1}{4}$

## 27:1

**1** 7 **2** 2 **3** 300 **4** 560 **5** 12 **6** 27 **7** 60 **8** 7200
**9** 9·41 **10** 83 456·9 **11** a 893 mm b 8·345 km c 297·8 cm
d 7200 m e 3750 m **12** a 66 cm b 28 cm **13** a B b B
**14** a $\frac{5}{6}$ b $\frac{5}{12}$ **15** a 1·53 b 17 476·2 c 8·1346

## 27:2

**1** 1080 **2** 519 **3** 32 **4** 45 L **5** 6 **6** 52 **7** 60 **8** 13
**9** 17·19 **10** 29 076·8 **11** 16 **12** 750 mL
**13** a $\frac{17}{100}$ b $\frac{9}{100}$ c $\frac{11}{100}$ d $\frac{89}{100}$ **14** Area = 55 $cm^2$, Perimeter = 36 cm
**15** a $\frac{3}{5}$ (Answers may vary) b $\frac{2}{8}$ (Answers may vary)
c $\frac{1}{3}$ (Answers may vary) **16** a 15 b 4 c 8 d 16 **17** 829 m

### Activity

a 0·5 b 0·6 c 0·35 d 0·52 e 0·875 f 0·0625 g 0·625 h 0·1875
i 0·28 j 0·03125 k 0·015625 l 0·984375

## 27:3

**1** 4755·97 **2** 277 217 **3** Area = 88 $m^2$, Perimeter = 46 m
**4** a $\frac{63}{100}$ b $\frac{43}{100}$ c $\frac{21}{100}$ d $\frac{47}{100}$ **5** a 21 b 63 c 14
**6** 1, 42, 2, 21, 3, 14, 6, 7
**7** a $\frac{4}{10}$ (Answers may vary) b $\frac{8}{10}$ (Answers may vary)
c $\frac{2}{10}$ (Answers may vary) **8** 5 oranges **9** \$20

## 27:4

**1** a 4·5 $cm^2$ b 4·5 $cm^2$ **2** 6 cm **3** Peter **4** 4 pizzas
**5** a 900 b 90 000

### Challenge

| $\frac{1}{2}$ | | | | $\frac{1}{2}$ | | | |
|---|---|---|---|---|---|---|---|
| $\frac{1}{4}$ | | $\frac{1}{4}$ | | $\frac{1}{4}$ | | $\frac{1}{4}$ | |
| $\frac{1}{8}$ | $\frac{1}{8}$ | $\frac{1}{8}$ | $\frac{1}{8}$ | $\frac{1}{8}$ | $\frac{1}{8}$ | $\frac{1}{8}$ | $\frac{1}{8}$ |

### Activity

8, 32, 48, 24, 40, 64, 16, 72, 56, 80
6, 24, 36, 18, 30, 48, 12, 54, 42, 60

## 28:1

**1** 56 **2** 511 **3** 827 **4** 3500 **5** $\frac{5}{12}$ **6** 55 **7** 320
**8** 6 **9** 44 334 **10** 14·73 **11** 20 mm **12** a $\frac{4}{5}$ (Answers may vary)
b $\frac{2}{5}$ (Answers may vary) c $\frac{3}{5}$ (Answers may vary)
**13** a 69 mm b 2·947 km c 545·6 cm d 3700 m e 1400 m
**14** 30 min **15** 16:28 **16** a 5 b 14 c 15 d 9

## 28:2

**1** 213 **2** 171 **3** 14 **4** $\frac{5}{10}$ or $\frac{1}{2}$ **5** 20 **6** 52 **7** 4800 L
**8** 12 **9** 56 866 **10** 9·44 **11** a 4·4 m b 13·2 m
**12** a $\frac{4}{5}$ b $\frac{4}{8}$ or $\frac{1}{2}$ **13** $2\frac{3}{4}$ **14** a $\frac{3}{4}$ b $\frac{3}{4}$ **15** a $\frac{2}{5}$ b $\frac{6}{18}$ **16** 9 h
**17** $\frac{2}{3} - \frac{1}{6} = \frac{4}{6} - \frac{1}{6} = \frac{3}{6}$ **18** a 4:35 pm b 4:09 am

### Activity

(1) millimetres (2) centimetres (3) metres (4) kilometres (24) minutes (25) hours (26) before noon (27) after noon (28) 5:40 am (29) 19:30

## 28:3

**1** 5772233 **2** 111229 **3** $\frac{3}{10}$ **4** 1, 32, 2, 16, 4, 8

**5** 9, 18, 27, 36, 45, 54, 63, 72, 81, 90 **6** **a** 25 **b** 49

**7** **a** $\frac{5}{8}$ **b** $\frac{5}{6}$ **c** $\frac{7}{10}$ **d** $\frac{5}{12}$

**8** **a** $\frac{2}{12}$ (Answers may vary) **b** $\frac{1}{4}$ (Answers may vary)

**c** $\frac{2}{5}$ (Answers may vary) **d** $\frac{1}{1}$ (Answers may vary)

**e** $\frac{2}{6}$ (Answers may vary) **f** $\frac{8}{10}$ (Answers may vary)

**9** 0·008 or $\frac{8}{1000}$ **10** **a** 18 min **b** 27 min **c** 70 min or 1 h 10 min

## 28:4

**1** **a** 1963 **b** 1942 **2** Area = 95 $m^2$, Perimeter = 58 m **3** 6 L

**4** □ = 14, △ = 4

**5** **a** 65·95 kg **b** 34·05 kg **6** **a** 279 **b** 12

### Challenge

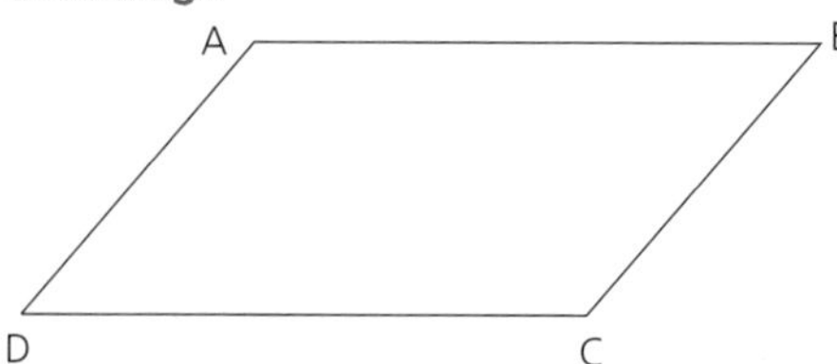


It is a quadrilateral. It has two pairs of parallel sides that are equal in length. Its opposite angles are equal in size.

### Activity

**a** 3

**b** kangaroos: 12, cows: 19

## 29:1

**1** 63 **2** 560 **3** 45 **4** 22 **5** 1 and 17. **6** 4 **7** 45 **8** $\frac{1}{8}$

**9** 285677 r 1 **10** 12·25

**11** **a** $\frac{4}{10}$ (Answers may vary) **b** $\frac{2}{4}$ (Answers may vary)

**c** $\frac{6}{8}$ (Answers may vary) **12** 10 **13** $1\frac{2}{4}$ or $1\frac{1}{2}$

**14** **a** 40000 **b** 50 **c** 0·008 **15** **a** 35679 **b** 874532

**16** **a** $\frac{2}{4}$ or $\frac{1}{2}$ **b** $\frac{3}{4}$ **17** 34 min **18** 12

## 29:2

**1** 168 **2** 27 L **3** 17 **4** 34 **5** $\frac{10}{12}$ or $\frac{5}{6}$ **6** 252 **7** 4900

**8** $6 **9** 40811 r 1 **10** 3·19 **11** **a** $\frac{1}{2} < \frac{4}{6}$ **b** $\frac{3}{4} > \frac{3}{8}$

**12** $15 **13** 26 **14** 17:26

**15** **a** **b**

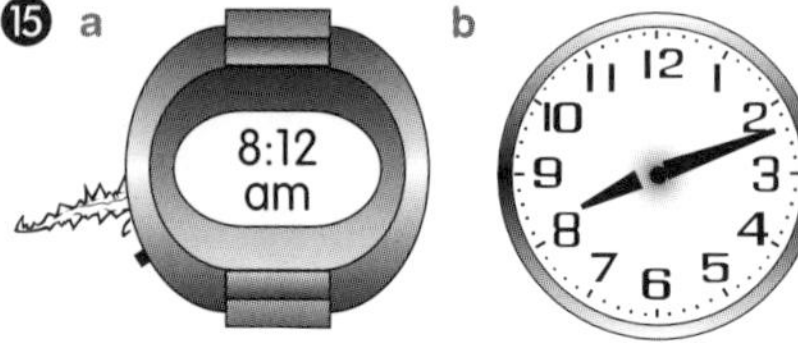


**16** $\frac{3}{10} + \frac{1}{2} = \frac{3}{10} + \frac{5}{10} = \frac{8}{10}$ **17** **a** 7:27 pm **b** 4:09 am **18** 225·6 cm

**19** 48 **20** 81 $m^2$ **21** 66

### Activity

**a** 20 m per second **b** 48 km per hour **c** 245 km per day **d** 12 mm per second

## 29:3

**1** 9883055 **2** 406151 **3** 17:26

**4** **a** $\frac{18}{20}$ (Answers may vary) **b** $\frac{2}{6}$ (Answers may vary)

**c** $\frac{1}{1}$ (Answers may vary) **5** **a** 2 am **b** 6:39 am

**6** $\frac{3}{6} + \frac{1}{3} = \frac{3}{6} + \frac{2}{6} = \frac{5}{6}$

**7** **a** **b**

**8** **a** kite **b** yes **9** $28 **10** **a** $\frac{7}{8}$ **b** $\frac{1}{8}$

**11** **a** 3·52 $m^2$ **b** 8 m 60 cm or 8·6 m

## 29:4

**1** **a** 18 **b** 27 **2** **a** 2040 **b** 840 **c** 1080 **d** 880

**3** **a** 0·6 m or 60 cm **b** 5·4 $m^2$ **4** Ty

### Challenge

Answers may vary. E.g. It is a 9-digit odd number. It becomes 356000000 when rounded to the nearest million. 712934906 is double this number.

### Activity

Put a tick for YES ✓

Put a cross for NO ✗

| | Drama | Gym | Tennis | Monday | Tuesday | Friday |
|---|---|---|---|---|---|---|
| Heather | ✗ | ✓ | ✗ | ✓ | ✗ | ✗ |
| Naomi | ✓ | ✗ | ✗ | ✗ | ✓ | ✗ |
| Luke | ✗ | ✗ | ✓ | ✗ | ✗ | ✓ |

| Answers | Club | Day |
|---|---|---|
| Heather | Gym | Monday |
| Naomi | Drama | Tuesday |
| Luke | Tennis | Friday |

## 30:1

**1** 787 **2** 614 **3** 75 **4** $\frac{5}{12}$ **5** 20 **6** $\frac{7}{10}$ **7** 21

**8** 650 cm **9** **a** 20 **b** 8 **c** 7 **d** 21 **10** **a** 2:24 am **b** 12:56 pm

**11** **a** 30 cm **b** 56 $cm^2$ **12** **a** 24 $m^2$ **b** 28 m **13** 14:30 **14** $28

**15** 72 **16** **a** 296 mm **b** 2·98 km

## 30:2

**1** 17 **2** 104 **3** 540 L **4** $\frac{5}{6}$ **5** 553566 **6** 12·5 **7** 40

**8** 15 **9** 40 **10** 840975 **11** **a** 08:50 **b** 20:50

**12** $\frac{4}{5} - \frac{2}{10} = \frac{4}{5} - \frac{1}{5} = \frac{3}{5}$ or $\frac{4}{5} - \frac{2}{10} = \frac{8}{10} - \frac{2}{10} = \frac{6}{10}$ **13** 40%

**14** **a** 64 m **b** 202 $m^2$ **15** **a** 90 min **b** 135 min

**16** **a** $\frac{1}{2} + \frac{1}{3} = \frac{3}{6} + \frac{2}{6} = \frac{5}{6}$

**b** $\frac{3}{4} + \frac{1}{5} = \frac{15}{20} + \frac{4}{20} = \frac{19}{20}$

**17** 7·25 h



  • ISBN 978 0 6557 0886 5

**Activity**

**1** a 25 b 13 c 3 **2** a 15 b 6

## 30:3

**1** 5235363 **2** 75968 **3** \$1 **4** 357·2 m **5** a $\frac{1}{10} < \frac{4}{5}$ b $\frac{5}{6} > \frac{2}{3}$
**6** 1 h 27 min or 87 min **7** 1156 m² **8** 31
**9** a $\frac{3}{5} - \frac{1}{2} = \frac{6}{10} - \frac{5}{10} = \frac{1}{10}$ b $\frac{3}{4} - \frac{1}{3} = \frac{9}{12} - \frac{4}{12} = \frac{5}{12}$
**10** a −0·05 b 0·05 c −0·2 **11** a 4:51 pm b 7:36 am **12** a 34 mm b 8·937 km c 84·5 cm d 2800 m

## 30:4

**1** $\frac{5}{12}$ **2** \$3302 **3** 12 **4** \$801 **5** My number is 738.
**6** a 64 m b 84 m²

**Challenge**

Answers will vary. See 30:4, Question 5 for an example.

**Activity**

Answers may vary. True is =384, False is ?17\$4

## 31:1

**1** 6 **2** 12 **3** 3 **4** 30 **5** 5039·9 **6** $\frac{5}{8}$ **7** 97 **8** 8 **9** 33
**10** 331·94 **11** 54 m² **12** A **13** a 12 m² b 42 m²
**14** a $\frac{1}{5} + \frac{4}{10} = \frac{1}{5} + \frac{2}{5} = \frac{3}{5}$ b $\frac{3}{6} + \frac{2}{12} = \frac{3}{6} + \frac{1}{6} = \frac{4}{6}$ or $\frac{2}{3}$
**15** a 3000 mL b 9261 mL c 2734 mL

## 31:2

**1** 10 **2** 4 **3** 700 **4** 6 **5** 250 **6** 13 **7** 55 **8** 273
**9** 71291 **10** 12·23 **11** 6 **12** 15
**13** a $\frac{8}{10} - \frac{1}{2} = \frac{8}{10} - \frac{5}{10} = \frac{3}{10}$ b $\frac{5}{7} + \frac{2}{10} = \frac{50}{70} + \frac{14}{70} = \frac{64}{70} = \frac{32}{35}$
**14** a 0·386 L b 0·946 L **15** a $10\frac{1}{2}$ b $10\frac{1}{2}$
**16** \$10 **17** \$104 **18** 3 h 21 min **19** 30 km

**Activity**

a 25 b 14 c 33 d 0 e 24 f 10 g 12 h 26 i 15 j 3 k 40

## 31:3

**1** 21164 **2** 4·32 **3** 175 min **4** 2068
**5** a $\frac{9}{16} - \frac{1}{8} = \frac{9}{16} - \frac{2}{16} = \frac{7}{16}$ b $\frac{1}{12} + \frac{1}{3} = \frac{1}{12} + \frac{4}{12} = \frac{5}{12}$
**6** a 8 L b 9·465 L c 7·234 L **7** a $\frac{1}{5}$ b $\frac{9}{15}$
**8** a 1200 b 3600 c 15000 **9** a 8000 mL b 4·839 L
**10** 8·5 or $8\frac{1}{2}$ **11** 24

## 31:4

**1**

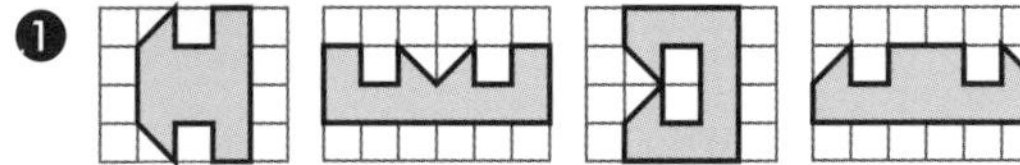

**2** 21 h **3** Answers will vary. **4** 22 seconds

**Challenge**

Answers will vary.

**Activity**

a 3 m b 3 m c 3.8 m d 13.6 m

## 32:1

**1** 37 **2** 164 **3** 2400 **4** 39 **5** 10313·7 **6** $\frac{1}{10}$ **7** 522
**8** 3 **9** 49 **10** 687·14 **11** 15 **12** $0·7 = \frac{70}{100} = 70\%$ **13** 48
**14** a $\frac{2}{10} + \frac{3}{4} = \frac{4}{20} + \frac{15}{20} = \frac{19}{20}$ b $\frac{3}{4} + \frac{1}{3} = \frac{9}{12} + \frac{4}{12} = \frac{13}{12}$
**15** a 6846 mL b 7278 mL c 8 kL **16** a 490 b 49000 c 350000

## 32:2

**1** 300 L **2** 7 **3** 980 **4** 9

**5**
```
   3 5
 × 7 6
 2 1 0   (6 × 35)
2 4 5 0  (70 × 35)
2 6 6 0
```
**6** \$4 **7** 17400 **8** 46 **9** 365

**10**
```
   9 3
 × 6 7
 6 5 1   (7 × 93)
5 5 8 0  (60 × 93)
6 2 3 1
```
**11** 16 times **12** 4935

**13** a $\frac{6}{7} - \frac{3}{10} = \frac{60}{70} - \frac{21}{70} = \frac{39}{70}$ b $\frac{3}{10} + \frac{1}{3} = \frac{9}{30} + \frac{10}{30} = \frac{19}{30}$
**14** 7 cm² **15** a 0·846 L b 0·058 L

**Activity**

120, 60, 300, 480, 600, 240, 540, 360, 180, 420
140, 280, 350, 420, 700, 70, 630, 560, 210, 490
720, 400, 160, 320, 800, 640, 560, 240, 80, 480

## 32:3

**1** 6349 **2** 7·87

**3**
```
   2 3
 × 4 5
 1 1 5   (5 × 23)
 9 2 0   (40 × 23)
1 0 3 5
```
**4**
```
   8 6
 × 6 7
 6 0 2   (7 × 86)
5 1 6 0  (60 × 86)
5 7 6 2
```

**5** a $\frac{8}{12} - \frac{1}{4} = \frac{8}{12} - \frac{3}{12} = \frac{5}{12}$ b $\frac{3}{4} + \frac{1}{10} = \frac{15}{20} + \frac{2}{20} = \frac{17}{20}$
**6** a 0·298 L b 0·046 L c 0·007 L **7** \$1560
**8** a 2000 b 6000 c 10000 **9** a 15 kL b 4000 L c 76395 L d 3000 kL e 57·684 ML **10** 73

## 32:4

**1** a 200 b 2400 **2** \$72 **3** a 350 m b 2800 m or 2·8 km
**4** 90 **5** 7585

**Challenge**

Answers will vary. E.g. An Olympic swimming pool holds 2·5 ML.

**Activity**

16, 21, 14, 22, 15, 18, 17, 20, 19, 23
a 18 b 24 c 32

## 33:1

**1** 86 **2** 561 **3** 3000 **4** 6 **5** 536 **6** $\frac{22}{30}$
**7** \$8 **8** 6434 m **9** 2400 L **10** 378 **11** a 5926 mL b 2978 mL c 6 kL **12** 175 km **13** g and kg will be underlined.

**14** a kL b ML c t **15** a 6000 kg b 3 t

**16** a yes b yes c $\frac{7}{10}$ d $\frac{7}{10}$

## 33:2

**1** 860 **2** 540 **3** 62 **4** $9

**5**
```
   5 9
 × 4 6
  3 5 4   (6 × 59)
2 3 6 0   (40 × 59)
2 7 1 4
```
**6** $\frac{31}{30}$ **7** $13.27 **8** 377 **9** 435

**10**
```
   7 1
 × 5 3
  2 1 3   (3 × 71)
3 5 5 0   (50 × 71)
3 7 6 3
```
**11** 78 291 **12** a $180 b $290

**13** a 0·298 L b 0·084 L c 0·019 L

**14** a 7400 kg b 90 t **15** a 75 kg b 210 kg

**Activity**

72, 23, 50, 77, 61, 65, 39, 54, 28, 86

91, 28, 49, 73, 16, 35, 70, 82, 54, 67

## 33:3

**1**
```
   8 3
 × 7 6
  4 9 8   (6 × 83)
5 8 1 0   (70 × 83)
6 3 0 8
```
**2**
```
    9 2 6
 ×    7 8
  7 4 0 8   (8 × 926)
6 4 8 2 0   (70 × 926)
7 2 2 2 8
```
**3** a 140 t b 1800 kg **4** a 2970 kg b 30 t **5** a 14:44 b 15:02

**6** a 60 b 43 c 22 d 126 **7** a 0·289 L b 0·037 L c 0·006 L

**8** a 36 kL b 9000 L c 27 746 L d 7000 kL e 27·453 ML

## 33:4

**1** 5·75 t

**2**

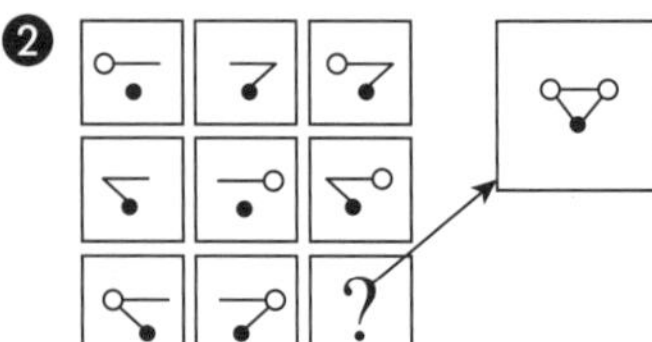

**3** a 100 b 20 **4** 8 times

**Challenge**

Answers will vary. E.g. Semitrailer: 50 t.

**Activity**

**1** a 96 b 160 c 610 d 204 e 66 f 36

**2** a LXXV b CCCLXXX c CCCLXXVIII d MMDLXVII e DCXXXV f MMMCMXCIX

## 34:1

**1** 2 **2** 12 **3** $\frac{21}{12}$ or $1\frac{9}{12}$ **4** 30 **5** 522 **6** $32

**7** 8475 m **8** 421 **9** $3.50 **10** 672

**11**

| Tonnes | Kilograms |
|---|---|
| 7 t | 7000 kg |
| 70 t | 70 000 kg |
| 7·546 t | 7546 kg |

| Litres | Millilitres |
|---|---|
| 9 L | 9000 mL |
| 2 L | 2000 mL |
| 8·3 L | 8300 mL |

**12** a $\frac{8}{10} > \frac{3}{5}$ b $\frac{2}{6} < \frac{2}{3}$ **13** 127 **14** mg is short for milligrams. **15** 6

**16** a kilograms (kg) b milligrams (mg) c grams (g)

**17** triangular pyramid

## 34:2

**1** 53 **2** 36 **3** 7 **4** $\frac{11}{12}$ **5** 3738 **6** $21.40 **7** 975 **8** 27

**9** 174 **10** 47 652 **11** 115·0 **12** 222·3 **13** 251·6 **14** a $20\frac{1}{2}$ b 11

**15**

| Tonnes & kilograms | Tonnes |
|---|---|
| 7 t 300 kg | 7·3 t |
| 9 t 546 kg | 9·546 t |
| 5 t 476 kg | 5·476 t |

**16** 9 **17** a 4·768 kg b 4000 mg c 67 000 kg d 46 800 g e 0·01 g

**Activity**

a 1 : 50 b 150 cm c 100 cm d Bed e Desk f J1 or J2

## 34:3

**1** 2380 **2** 1314 **3** 17 664 **4** 76·7 **5** 210·0 **6** 539·0

**7** a 18·2 t b 91 t **8** $\frac{1}{2} = \frac{2}{4} = \frac{4}{8} = \frac{5}{10}$ **9** a 9100 kg b 40 t

**10** 852 **11** 2·934 kg **12** 7 **13** a 4000 mg b 0·01 g **14** $12 000

**15**

| □ | 20 | 13 | 26 |
|---|---|---|---|
| △ | 920 | 598 | 1196 |

## 34:4

**1** 552 **2** 1 and 72, 2 and 36, 3 and 24, 4 and 18, 6 and 12, 8 and 9

**3** a 8·66 kg or 8660 g b 1·34 kg or 1340 g **4** 131·25 mg

**5** 37

**6** 8976

**Challenge**

a 9 b 85 c 49 d 63 e 13 f 8

**Activity**

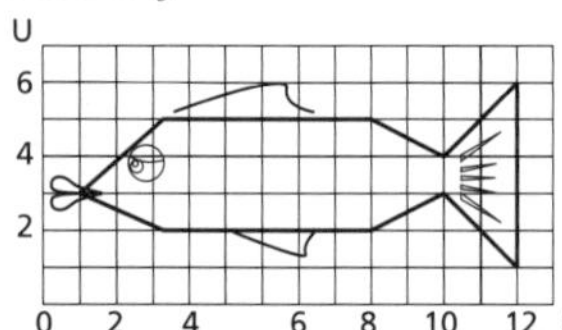


## 35:1

**1** 76 **2** 6 **3** 8 **4** $\frac{1}{4}$ **5** 522 **6** 77 **7** 5 **8** 5498 L

**9** $400 **10** 308 **11** 4 **12** a 7000 kg b 8 t c 276 g d 7925 g e 2 $km^2$ **13** □ = ___÷___, △ = ___+___, or vice versa

**14**

| Decagons | 1 | 2 | 3 | 4 | 5 |
|---|---|---|---|---|---|
| Sides | 10 | 20 | 30 | 40 | 50 |

**15** a $\frac{4}{5}$ b $\frac{3}{12}$

**16** a A prime number has 2 factors, 1 and itself.
b A composite number has more than 2 factors. **17** 16

**18** a ha b $km^2$

## 35:2

**1** 1099 **2** 386 **3** $\frac{26}{30}$ **4** $1\frac{3}{5}$ **5** 2849 **6** 18 **7** 12·5 **8** 35

❾ $\frac{1}{6}$ ❿ 15 402 ⓫ 42·0 ⓬ 509·2 ⓭ 660·0

⓮ **a** 7760 kg **b** 20 t **c** 456 g **d** 2487 g **e** 9000 mg **f** 0·01 g

⓯ 0·061 kg or 61 g ⓰ 2 ⓱ 17, 29, 83, and 67 will be circled.

⓲ **a** 2 km² **b** 6 ha **c** 5 km² **d** 900 ha

**Activity**

**a** =384 (True) **b** =384 (True) **c** =384 (True)

## 35:3

❶ 2117 ❷ 205·2 ❸ 8763 ❹ $\frac{3}{4}+\frac{1}{8}=\frac{6}{8}+\frac{1}{8}=\frac{7}{8}$

❺ **a** 4340 kg **b** 50 t **c** 907 g **d** 5264 g **e** 4000 mg

❻ **a** 37 **b** 60 **c** 124 ❼ **a** 12·6 t **b** 26·6 t ❽ **a** 2·4 t **b** 4 t

❾ **a** 2 ha **b** 4 ha

## 35:4

❶ **a** 147 943 ML **b** 751 022 ML ❷ 335

❸ **a**

```
      6 · [9][2] 7
+  [8] · 0  4 [9]
 ----------------
  1 4 · 9  7  6
```

**b**

```
      8 · [7] 5 [2]
+  [7] · 8 [7]  8
 ----------------
  1 6 · 6  3  0
```

❹ 36 mg

**Challenge**

Answers will vary. E.g. A grain of rice weighs 21 mg.

**Activity**

Factors: 1, 3, 5, and 15. Two rectangles of 1 x 15 and 3 x 5 will be drawn.

## 36:1

❶ 765 ❷ 543 ❸ $3600 ❹ 8 ❺ 1233·35 ❻ 1 ❼ 80

❽ 10 ❾ $\frac{1}{10}$ ❿ $8808·01 ⓫ 9, 12, 6, and 15 will be circled.

⓬ **a** $\frac{3}{5}$ **b** $\frac{3}{24}$

⓭ 2, 3, 5 and 7 will be ticked.

| | | | |
|---|---|---|---|
| 1 | 1 | | |
| 2 | 1 | 2 | |
| 3 | 1 | 3 | |
| 4 | 1 | 2 | 4 |

| | | | | |
|---|---|---|---|---|
| 5 | 1 | 5 | | |
| 6 | 1 | 2 | 3 | 6 |
| 7 | 1 | 7 | | |
| 8 | 1 | 2 | 4 | 8 |

⓮ **a** 21 **b** 4 **c** 12 **d** 20 ⓯ **a** A, C, and E. **b** B and D.

## 36:2

❶ 84 ❷ 68 ❸ 24 ❹ $\frac{4}{30}$ or $\frac{2}{15}$ ❺ 43·3333 ❻ 100 ❼ 127

❽ $5\frac{1}{6}$ ❾ 5624 m ❿ 853·108 ⓫ **a** 7 km² **b** 2 ha **c** 9 km² **d** 500 ha ⓬ 4 ⓭ $\frac{7}{10}-\frac{1}{5}=\frac{7}{10}-\frac{2}{10}=\frac{5}{10}$ or $\frac{1}{2}$

⓮ 790, 685, and 917 080 will be circled.

⓯ 19, 31, and 47 will be circled.

⓰ 1, 42, 2, 21, 3, 14, 6, 7 ⓱ **a** 2 **b** 1

**Activity**

**a** Answers will vary.

**b**

| | a | e | i | o | u |
|---|---|---|---|---|---|
| Tally | 𝍸 I | 𝍸 𝍸 III | 𝍸 𝍸 I | 𝍸 𝍸 II | 𝍸 I |

**c** e

## 36:3

❶ 6255·84 ❷ $6021.32 ❸ **a** 10 min **b** 1 h 40 min

❹ 65 ❺ 50 ❻ 4·4 cm ❼ 2, 3, 5, 7, 11, 13, 17, 19, 23, and 29.

❽ **a** 8 km² **b** 4 ha **c** 9 km² **d** 1200 ha

❾ 11 and 13 will be ticked.

| | | | | | | |
|---|---|---|---|---|---|---|
| 9 | 1 | 9 | 3 | | | |
| 10 | 1 | 10 | 2 | 5 | | |
| 11 | 1 | 11 | | | | |
| 12 | 1 | 12 | 2 | 6 | 3 | 4 |

| | | | | | |
|---|---|---|---|---|---|
| 13 | 1 | 13 | | | |
| 14 | 1 | 14 | 2 | 7 | |
| 15 | 1 | 15 | 3 | 5 | |
| 16 | 1 | 16 | 2 | 8 | 4 |

❿ 800 010, 8310, and 7850 will be circled.

⓫ 1, 100, 2, 50, 4, 25, 5, 20, 10

## 36:4

❶ **a**

```
        [7 5 9]
×         [3]8
 ------------
      6 0 7 2
    2 2 7 7 0
 ------------
   [2 8 8 4 2]
```

**b**

```
   [424]
×     54
 -------
   1696
  21200
 -------
 [22896]
```

**c**

```
   [296]
×     39
 -------
   2664
   8880
 -------
 [11544]
```

❷ 29 644

❸

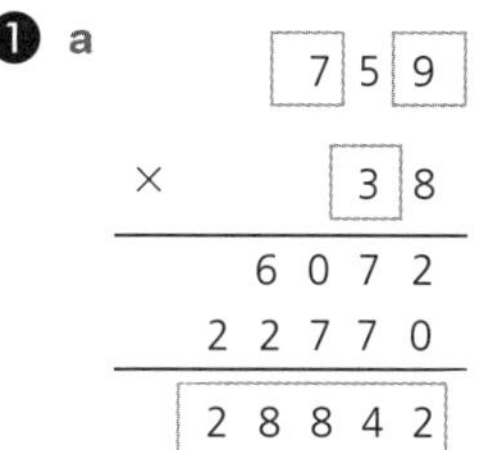

❹ 6·73 cm ❺ **a** 72 **b** 528 ❻ 905 ❼ 140

**Challenge**

Answers will vary.

**Activity**

**a** yes **b** yes **c** yes

## 37:1

❶ 0·8 ❷ 16 ❸ 12 ❹ $\frac{7}{8}$ ❺ 316 022 ❻ $4.20 ❼ 0·1

❽ 8345 ❾ 43 ❿ 452 513 ⓫ 6578 and 2080 will be circled.

⓬ **a** Eggs **b** $\frac{1}{4}$ **c** 2 **d** Bacon **e** Pastry ⓭ **a** 12 **b** 21 ⓮ 48 km

⓯ 0·003 or $\frac{3}{1000}$

## 37:2

❶ 3·3 ❷ 3 ❸ 17 ❹ $8\frac{3}{8}$ ❺ $1192.49 ❻ 24 ❼ 357 ❽ 58

❾ $\frac{23}{10}$ or $2\frac{3}{10}$ ❿ $6563.07 ⓫ 1, 80, 2, 40, 4, 20, 5, 16, 8, 10

⓬ **a** 122 **b** Male **c** 398 ⓭ 36, 100, 812, and 4000 will be circled.

⓮ 56 400

**Activity**

**A** (2, 2) **B** (−2, 1) **C** (−2, −2) **D** (0, −1) **E** (2, 0) **F** (0, 0)

Answers may vary. E.g. Triangle: (2, 2), (2, 0), (0, 0).

## 37:3

1 3648 2 792 054 3 56 404

4

| Number of horses | 1 | 2 | 3 | 4 | 5 |
|---|---|---|---|---|---|
| Number of legs | 4 | 8 | 12 | 16 | 20 |

**a** number of horses × 4 = number of legs **b** 2368 5 **a** 3 **b** 20

6 **a** 10 **b** 2 7 2307, 800 010, 410 043, and 8310 will be circled.

8 2 and 4.

## 37:4

1 **a** 525 **b** 869 **c** 78·96 2 Svetla 3 **a** 30% **b** 20%

4 60

**Challenge**

Answers will vary.

## 18:3 ☐ out of 7

1.
```
  8456735
    98365
  5260831
+   25498
```

2.
```
  $9000.00
– $8675.40
```

3. If this spinner is spun, which result, **Y**, **B** or **G**:
   **a** is most likely? ______
   **b** has a 25% chance of occurring? ______

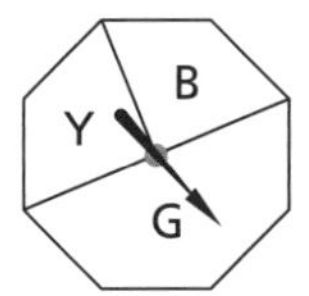


4. How far did I travel if I drove for 5 hours at 56 km/h? ______

5. Find the change from $50 if the total $35.68 is rounded off to the nearest 5 cents. ______

6. This table shows the goals Felicity and Lachlan scored in the first 12 weeks of soccer.

| Felicity | 0 | 1 | 6 | 1 | 5 | 0 | 4 | 0 | 0 | 0 | 0 | 1 |
|---|---|---|---|---|---|---|---|---|---|---|---|---|
| Lachlan | 3 | 6 | 8 | 5 | 0 | 1 | 0 | 1 | 2 | 3 | 1 | 11 |

How many goals were scored by:
**a** Felicity? ______ **b** Lachlan? ______
**c** both Felicity and Lachlan? ______
What was the mode for:
**d** Felicity? ______ **e** Lachlan? ______
What was the range for:
**f** Felicity? ______ **g** Lachlan? ______
What was the median for:
**h** Felicity? ______ **i** Lachlan? ______

7. Describe the spread of these scores.
______
______
______

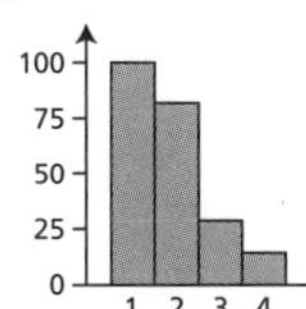


## 18:4 ☐ out of 5 Extension

1. The number of corners plus the number of faces minus the number of edges for 8 hexagonal prisms. ______

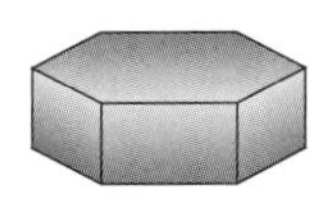

2. This cylinder is 76 cm tall. How high is:
   **a** one slice? ______
   **b** three slices? ______

3. One plane flew 12 km per minute for 7 hours. Another plane flew 13 km/min for 9 hours. What was the difference between the total distance they flew? ______

4. On a trip to the Northern Territory, our average speed was 67 km/h for the first part of the trip, 93 km/h for the second part and 85 km/h for the third part.
If we travelled for 3 hours in the first part of the trip, 2 hours in the second part and $4\frac{1}{2}$ hours in the third part, how far did we travel in total? ______

5. **a** CLVI ______ **b** CCCLXXXVI ______
   **c** CXVI ______ **d** CCLXXVII ______

**Challenge**

*Make a list of events and estimate the probability of each event, as a percentage.*
______
______
______
______

______
______
______
______
______

## 19:1   ☐ out of 13

1. 7 × 30 ______
2. 8 × 400 ______
3. 52 − 4 × 5 ______
4. 42 ÷ 7 ______
5. 0·6 + 0·25
6. Divide 35 by 7. ______
7. 0·2 + 0·5 ______
8. 0·3 + 0·6 ______
9. 83 + ______ = 106
10. 5·87 + 0·73

11.  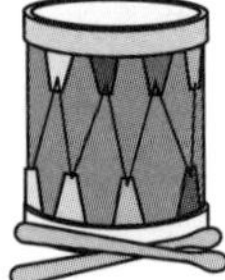 

$4.20   $7.60   $1.90

**a** Circle the best estimate for the cost of 3 ring toys.

$3   $6   $8   $11

**b** If each item were reduced by fifty cents, what is the total cost of a drum and a toy horse?

______

**c** Circle the best estimate for the cost of 2 drums.

$10   $13   $15   $18

12. | 5, 8, 3, 2, 2, 2, 4, 6 |
|---|

Write these scores in order from smallest to largest. ______

Find the:

**a** range ______   **b** median ______

**c** mode ______   **d** average ______

13. Arrange from smallest to largest.

67·25   54   7·21   8·9

______

## 19:2   ☐ out of 15

1. $57 × 8 ______
2. 4 × $390 ______
3. 670 − 174 ______
4. 280 ÷ 7 ______
5. 97·7 + 3·819
6. $\frac{4}{5}$ of 25 ______
7. 80% of 10 ______
8. 80% of 40 ______
9. 57 + ______ = 83
10. 4·5 billion − 2·3 billion

11. Find the total of these amounts, rounded off to the nearest 5 cents. ______

| $45.00<br>$83.00<br>$35.09 |
|---|

12. How far did I travel if I drove for $3\frac{1}{2}$ hours at 96 km/h? ______

13. For these scores, what is:

| 5, 6, 6, 7,<br>9, 13, 18 |
|---|

**a** the range? ______

**b** the mode? ______

**c** the median? ______

**d** the average score? ______

14. **a** 0·7 + 0·3 = ______   **b** 1·2 + 0·8 = ______

15. **Tally of chores done**

| | |
|---|---|
| Setting table | 𝍸 𝍸 |
| Clearing table | 𝍸 \| |
| Sweeping floor | 𝍸 |
| Wiping benches | 𝍸 𝍸 \|\|\|\| |

Isabella kept a tally of the jobs she did during the week.

**a** Which job did she do 14 times? ______

**b** How many jobs were recorded altogether? ______

**Eliminating possibilities**

How old is Alan if:

- reversing the digits of Naomi's age gives Alan's age,
- the total of their ages is 66,
- Naomi is under 20 years of age.

| NAOMI | ALAN |
|---|---|
| ~~19~~ | ~~91~~ |

Alan's age = ______

## 19:3 ☐ out of 6

1.
```
  7·867
+ 5·17
```

2.
```
  7·465 km
+ 3·978 km
```

3. Gillian began with this amount. She bought these items.

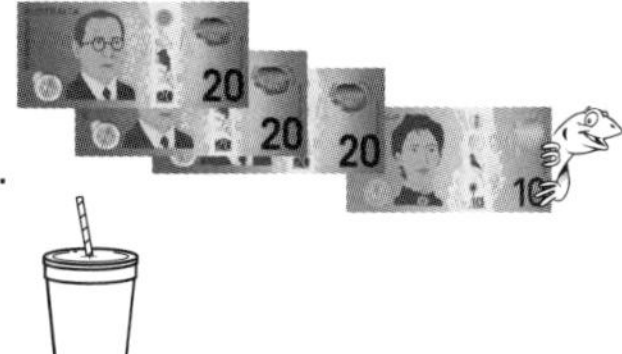

$15 | $35.40 | $1.80

Total amount spent: _______ Amount left: _______

4. For these scores, what is:

| 35, 36, 36, 37, 38, 38, 40, 40, 40, 47 |
|---|

a the range? _______
b the mode? _______
c the median? _______
d the average score? _______

5. Caleb sold three houses for $2·8 million, $3·5 million and $3·9 million. Write the total of these sales, expressed as millions. _______

6. Complete the table and graph

| Cups of water we drank | | Number |
|---|---|---|
| Jasmine | 𝍸 IIII | |
| Matilda | 𝍸 𝍸 II | |
| Lydia | 𝍸 II | |
| Isaac | 𝍸 𝍸 𝍸 I | |

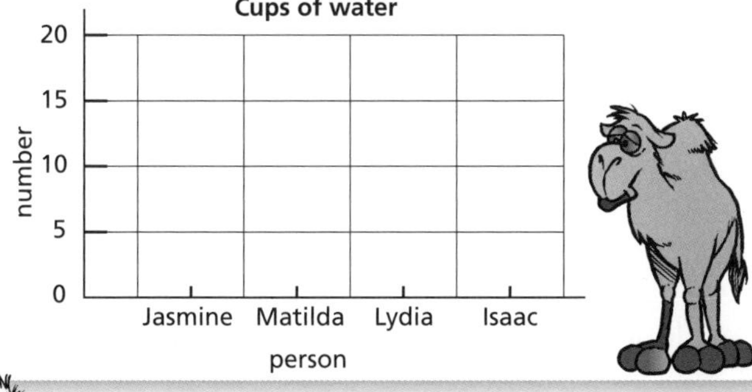


## 19:4 Extension ☐ out of 5

1. a
```
  7 ·☐7 5
+ ☐ · 8☐6
  17 · 1 8 1
```

2. b
```
  2 ·9 5☐
+ ☐ ·4☐ 6
  10 ·4 5 3
```

3. The number of faces plus the number of corners minus the number of edges on 24 square pyramids. _______

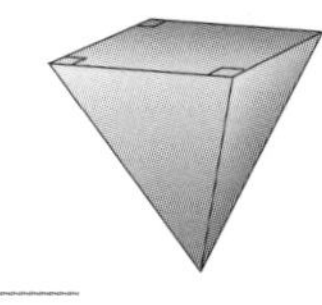

4. The top 5 countries in gold medals per ten million people in the 1996 Atlanta Olympics are shown.

| Country | Gold Medals ptmp* |
|---|---|
| Australia | 5·26 |
| Cuba | 6·36 |
| Denmark | 8 |
| New Zealand | 8·57 |
| Switzerland | 5·71 |

*ptmp = per ten million people

a Which country did the best? _______

b Australia's population was 19 million. How many gold medals did we win? _______

By how much did New Zealand beat:

c Denmark? _______ gold medals ptmp

d Australia? _______ gold medals ptmp

5. An 85-year-old Greek woman died in 180 BCE. When was she born? _______

### Challenge

*I surveyed all the members of my family and created this sector graph. How is this information misleading?*

**Australians who have visited Uluru**

yes
no

_______
_______
_______
_______

Consecutive numbers follow one another.

**4**, **5** and **6** are three **consecutive** numbers. Their sum is 15.

a On the list to the right, cross out those numbers that are the sum of 2 consecutive numbers, 3 consecutive numbers etc, up to the sum of 7 consecutive numbers.

b Which numbers are left? _______

What kind are they? _______

## 20:1 ☐ out of 17

1. $\frac{1}{4}$ of 20 ______
2. 50 ÷ 5 ______
3. 7 × ______ = 63
4. 45 ÷ 9 ______
5. $\begin{array}{r} 457.34 \\ +\ 35.75 \\ \hline \end{array}$
6. 56 + ______ = 100
7. 35 + 78 = ______ + 80
8. 25 × ______ = 50
9. 0·9 × 100 ______
10. $\begin{array}{r} 83.536 \\ +\ 9.946 \\ \hline \end{array}$
11. Lexi bought a house for $1·3 million and sold it for $6·5 million. What is the difference between the buying and selling price, expressed as millions? ______
12. Find the change from $30 if the total is $25. ______
13. a 8 km = ______ m
    b 7655 mL = ______ L
    c 45 mm = ______ cm
    d 356 cm = ______ m
14. I had one 7·6 m length of wood and one 4·8 m length of wood.
    a What is the total length of the wood? ______
    b What is the difference between the lengths of wood? ______
15. My mass is 28·56 kg and my bag is 3·67 kg. When I wear my bag, what is my total mass? ______
16. What is 4·5 grams less than 1 kg? ______
17. Arrange these in descending order:
    $5\frac{1}{4}$ $1\frac{3}{8}$ 4 $3\frac{1}{8}$ ______

## 20:2 ☐ out of 19

1. 520 − 147 ______
2. 846 − 297 ______
3. $\frac{3}{4}$ of 28 ______
4. 125 ÷ 5 ______
5. $\begin{array}{r} 286.478 \\ -\ 93.982 \\ \hline \end{array}$
6. $\frac{4}{10}$ of 50 ______
7. 50% of $36 ______
8. 8·5 cm × 1000 ______
9. 785 + 449 ______
10. $\begin{array}{r} 79.476 \\ +\ 8.637 \\ \hline \end{array}$
11. $9\overline{)639}$
12. $4\overline{)804}$
13. $9\overline{)734}$
14. Find the total of these amounts, rounded off to the nearest 5 cents. ______

15. I used 26·546 L of water on the garden and 35·756 L of water on the grass. What was my total water usage? ______
16. I surveyed my cricket team for this graph. Explain why this graph is misleading.
    ______
    ______

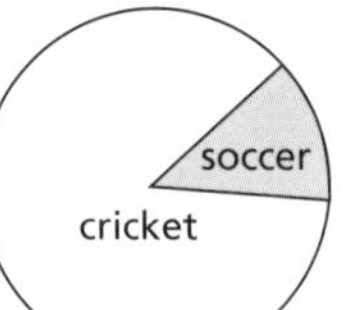


17. What is 1·4 metres less than a kilometre? ______

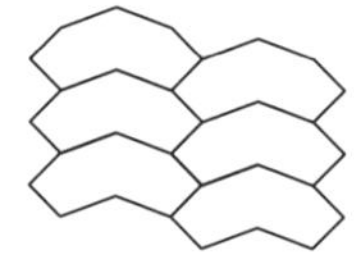

18. These shapes fit together without gaps or overlaps. They t ______.
19. I painted 5·78 m of my fence on the first day, 2·67 m on the next and 3·55 m to finish. How long is my fence? ______

ISBN 978 0 6557 0886 5

## 20:3 out of 9

1. $\begin{array}{r} 4{\cdot}086 \\ +\ 5{\cdot}789 \\ \hline \end{array}$

2. $\begin{array}{r} 3{\cdot}9988\text{km} \\ -\ 3{\cdot}8793\text{km} \\ \hline \end{array}$

3. On Monday I walked 4·867 km and on Tuesday I walked 5·347 km.
   a What was the total distance I walked? ____
   b What was the difference between the two walks? ____

4. a Which letter has a 50% chance that the spinner will land on it?

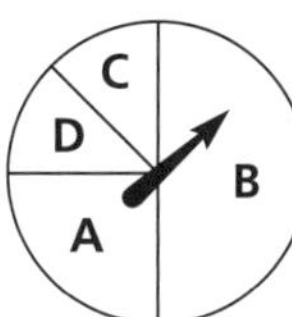


____

What is the chance, as a decimal, that the spinner will land on:

b **B**? ____ c **A**? ____

5. When wearing my bag, I weighed 33·75 kg. When I took my bag off, I weighed 29·87 kg.
   What was the mass of the bag? ____

6. 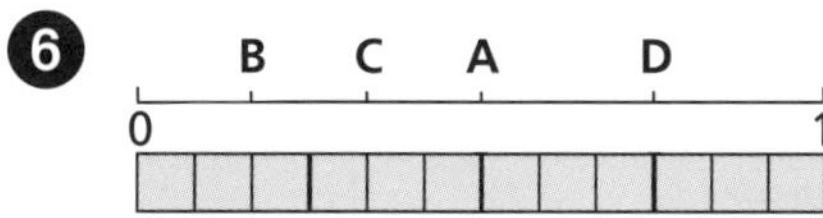


On this number line what fraction is at:

a **A**? ____ b **B**? ____

c **C**? ____ d **D**? ____

7. Explain what happened to A to make this shape?

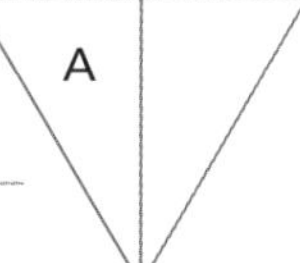


____

8. Create a pattern by flipping the tile in both directions.

9. $(3 \times 10^4) + (4 \times 10^3) + (9 \times 10^2) + (2 \times 10^1) + 7$
   = ____

## 20:4 Extension out of 7

1. 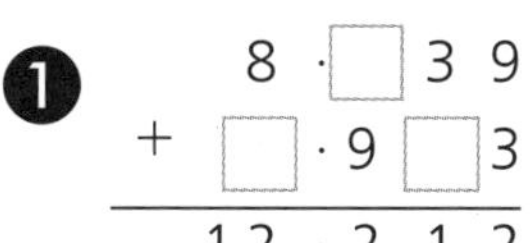


2. 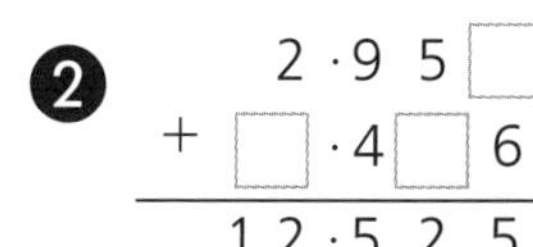


3. The number of faces plus the number of corners minus the number of edges on 26 square pyramids. ____

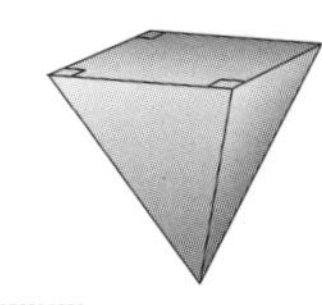

4. What is 4·545 kilograms less than 5 tonnes? ____

5. The shaded part has a value of 12. What is the value of the whole? ____

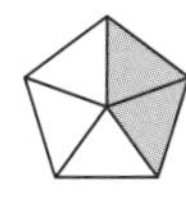

6. Jason has five times as many cards as Harvey and 35 less than Jen. How many cards do they have altogether if Harvey has 25? ____

7. I had 5·675 kg of flour in one bag and 3 kg 780 g in another. How much less than 10 kg did I have? ____

### Challenge

*Create a design on the grid by flipping, sliding and turning this pattern.*

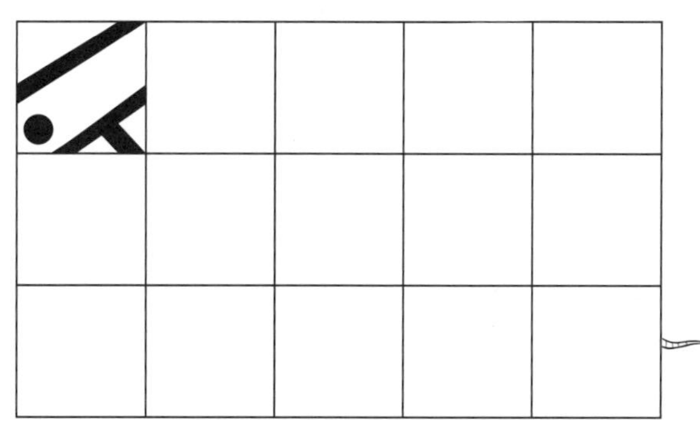

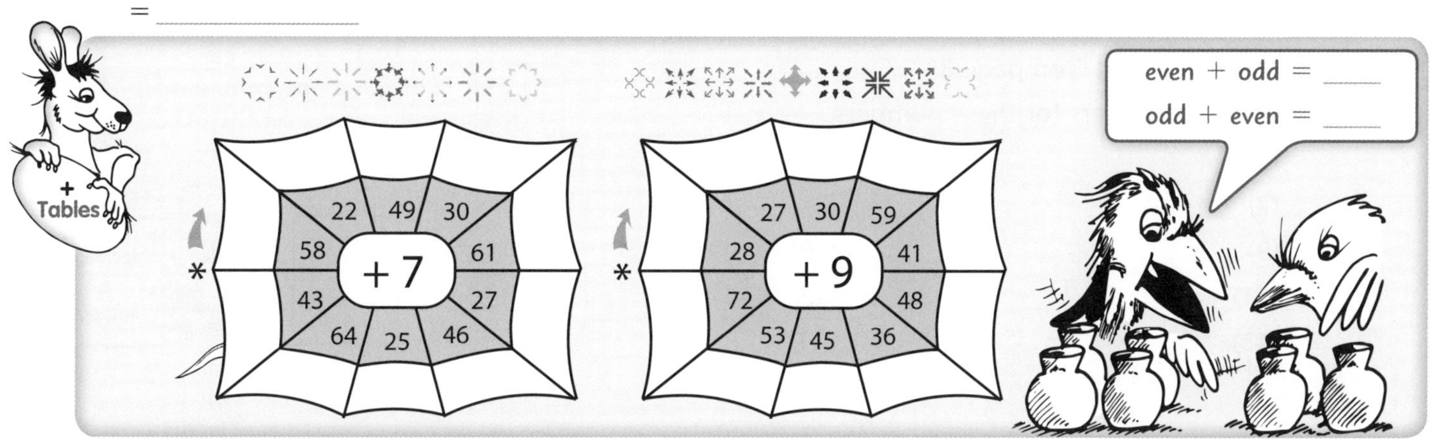

## 21:1 out of 17

1. 0·2 + 0·8 ______
2. $5 × 70 ______
3. 24 ÷ 8 ______
4. 32 ÷ 8 ______
5. 
```
  35·867
− 19·75
```
6. 100 m − 2·5 m ______
7. 8 × 800 ______
8. Zero squared. ______
9. $\frac{1}{3}$ of 36. ______
10. 
```
  36·87
+  8·93
```
11. What is 1·1 cm less than 1 m? ______
12. 

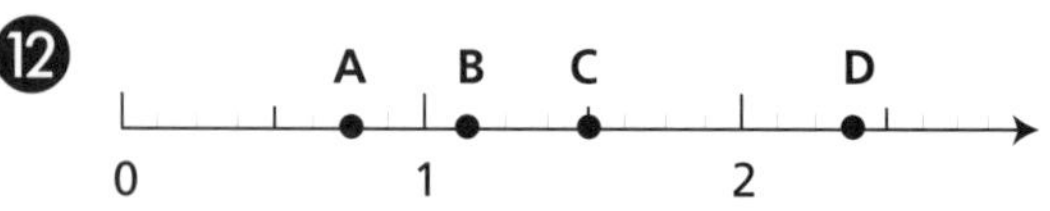


Give the fraction at the letter, as its simplest mixed numeral.

**a** **A** ______ **b** **B** ______

**c** **C** ______ **d** **D** ______

13. Write an estimate to the nearest whole number in the next algorithm, then answer both questions.

**a**
```
  3·367
+ 5·976   + ______
```
**b**
```
  6·204
+ 1·689   + ______
```

14. Round $56.89 to the nearest:

**a** 5 cents ______ **b** dollar ______

15. Round 9·864 to the nearest tenth. ______
16. **a** Halves in 5 wholes.

______

**b** Quarters in 5 wholes.

______

17. 6·546 kg = ______ g

## 21:2 out of 18

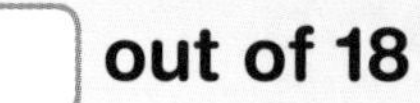

1. 720 ÷ 8 ______
2. $\frac{2}{5}$ of 15 ______
3. 700 ÷ 10 ______
4. $\frac{7}{10} - \frac{2}{5}$ ______
5. 
```
  $546·45
+ $356·78
```
6. $560 × 7 ______
7. Increase 34 by 35. ______
8. Triple 645. ______
9. 356 + ______ = 412
10. 
```
  $4782
×     7
```

11. $7\overline{)392}$
12. $9\overline{)576}$
13. $10\overline{)640}$
14. 

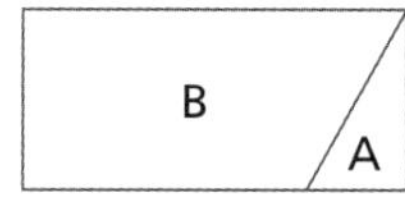


The first shape above is changed to make the second shape. What happened to A to change the shape?

______

______

15. What is 4·65 metres less than 20 metres? ______
16. Write an estimate, to the nearest whole number, in the next algorithm then answer both questions.

**a**
```
  9·465
+ 4·6     + ______
```
**b**
```
  7·364
+ 4·7     + ______
```

17. Round 45·326 to the nearest hundredth. ______
18. **a** Metres in 5·6 kilometres ______

**b** Kilograms in 6·9 tonnes ______

**c** Centimetres in 88 millimetres ______

**Turn to ID card C on page 8.**

**Give the answers for these numbers.**

| | |
|---|---|
| (5) ______ | (6) ______ |
| (7) ______ | (8) ______ |
| (9) ______ | (10) ______ |
| (11) ______ | (12) ______ |
| (13) ______ | (14) ______ |

Make a study card:
Put questions on one side and answers on the other.

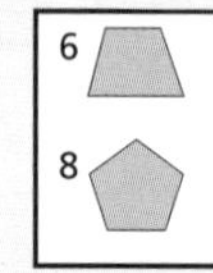

## 21:3

 out of 13

For Questions 1 and 2, write an estimate to the nearest whole number next to the algorithm, then answer both questions.

1. 
```
  3·198
+ 3·08      + ______
```

2. 
```
  4·579
+ 7·75      + ______
```

3. Arrange −2, 5 and − 3 in order from smallest to largest. ______

4. In a newspaper there are 50 pages. If $\frac{1}{10}$ of the newspaper contains comics, how many pages of comics are there? ______

5. Arrange in ascending order: 0·06, 0·63, 0·36 ______

6. How many seeds are in a packet if a quarter of the seeds can be used to plant 3 rows of 9 seeds? ______

7. 100 − 32·8 − 0·2 ______

8. Round 298·1057 to the nearest thousandth. ______

9. If you are facing north-west, what direction is to your:

   a left? ______ b right? ______

10. 18 − 9 + 5 − 12·15 ______

11. Round 32·6 to the nearest whole. ______

12. The temperature on Monday was – 5°C. What is the temperature if it was 7 degrees colder the following Monday? ______

13. How much of the block is covered:

   a as a fraction? ______

   b as a decimal? ______

## 21:4

**Extension**

out of 7

1. 
```
  □ · 7 0 □
+ 8 · □ 4 9
-----------
 1 3 · 1 6 1
```

2. 
```
  9 · 2 □ 7
+ □ · 9 6 □
-----------
 1 2 · 2 1 4
```

3. One seventh of the whole is 5. Write the value of the shaded part.

   a (3 of 7 parts shaded) ______ b (6 of 7 parts shaded) ______

4. A man born in 365 BCE died 40 years later. The year he died was ______.

5. Which one-digit number when multiplied by 7·79 gives an answer closest to 40? ______

6. The number of parallelograms of any size in this figure ______

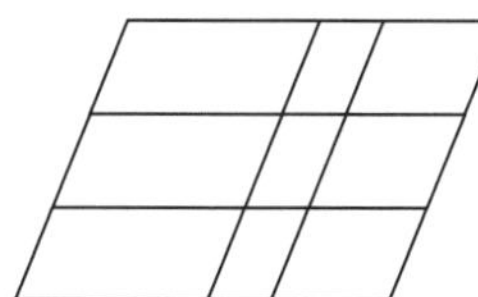

7. I ride in my car for 36·9 km every day, then ride my bike for 13·9 km. How far did I travel in:

   a a week? ______ b 6 weeks? ______

**Challenge**

*Draw your own tessellation.*

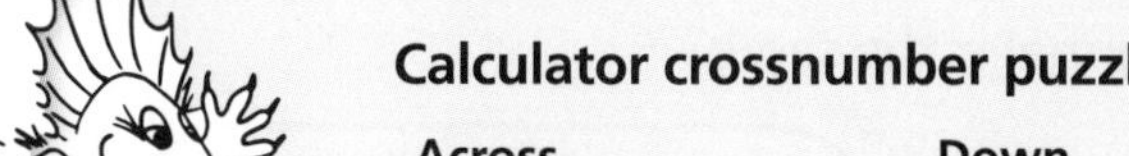

**Concept**

### Calculator crossnumber puzzle

**Across**

1 81 × 305
5 9743 − 9203
6 $6^2 + 4332$
8 475 ÷ 25
9 223 × 3
10 $821^2$

**Down**

1 4446 ÷ 19
2 9000 − 1631
3 275 ÷ 5
4 701 391 + 103 510
7 $56^2$
9 25 × 24

## 22:1

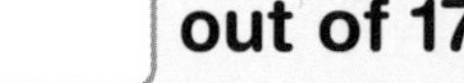

out of 17

1. 6·8 + 5 ______
2. 200 − 68 ______
3. 30 ÷ 6 × 6 ______
4. 4 squared ______
5. $\begin{array}{r} \$46·37 \\ +\ \$27·97 \\ \hline \end{array}$
6. 7 times 900. ______
7. $\frac{3}{5}$ of 40 ______
8. 0·7 × 100 ______
9. $\frac{3}{10} + \frac{4}{10}$ ______
10. $\begin{array}{r} \$45·9 \\ \times\ \ \ 3 \\ \hline \end{array}$

11. 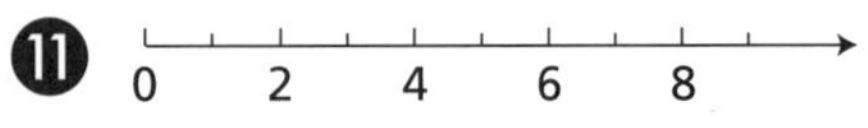


How many counting numbers are:

a 2 units away from 4? ______

b less than 3 units from 4? ______

12. 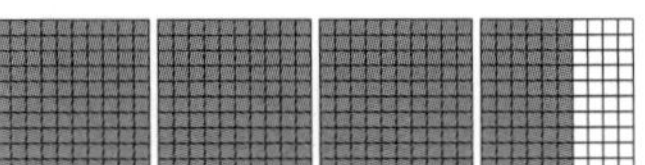

Show this number as a:

a decimal ______

b mixed number ______

13. Rhonda drinks 600 mL of juice each day. How many litres does she drink in a week? ______

14. List all the possible outcomes if I roll 2 standard dice and add the numbers shown.

______

15. Write as an improper fraction:

a $2\frac{1}{4}$ ______ b $3\frac{3}{5}$ ______

16. Arrange in descending order.

75 124 911 75 241 119 75 921 977

______

17. Complete the pattern:

56·36, 58·39, 60·42, ______, ______

## 22:2

out of 20

1. 99 × 6 ______
2. 400 ÷ 8 ______
3. $5^2 + 546$ ______
4. 560 ÷ 8 ______
5. $\begin{array}{r} \$26.57 \\ +\ \$74.89 \\ \hline \end{array}$
6. 6·7 km × 1000 ______
7. 8 ÷ (64 − 56) ______
8. $\frac{8}{12} + \frac{3}{4}$ ______
9. Triple 297. ______
10. $\begin{array}{r} 35·46 \\ \times\ \ \ \ 9 \\ \hline \end{array}$

11. $4\overline{)648}$ 12. $7\overline{)936}$ 13. $8\overline{)773}$

14. I used 0·8 m of ribbon to wrap each present. How many metres did I use for 8 presents? ______
15. What is the percentage probability of tossing a head when I toss a coin? ______
16. What is the difference between walking straight to school (12 072 paces) or going via the shops (16 180 paces)? ______
17. A stone is kicked 5·8 m twice, then 7·8 m, and finally 12·1 m twice. How far has it been kicked altogether? ______
18. List the factors of 18. ______
19. Write an estimate to the nearest whole number in the next algorithm, then answer both questions.

a $\begin{array}{r} 8·235 \\ +\ 1·856 \\ \hline \end{array}$ $\begin{array}{r} \\ + \\ \hline \end{array}$  b $\begin{array}{r} 7·079 \\ +\ 3·907 \\ \hline \end{array}$ $\begin{array}{r} \\ + \\ \hline \end{array}$

20. I bought 600 bananas at 75c each. What was the total cost? ______

In a magic square, the sum of each row, column and diagonal is the same.

| 11 | 6 | 7 |
|---|---|---|
| 4 | 8 | 12 |
| 9 | 10 | 5 |

Here, all lines add up to 24.

Complete these magic squares.

a

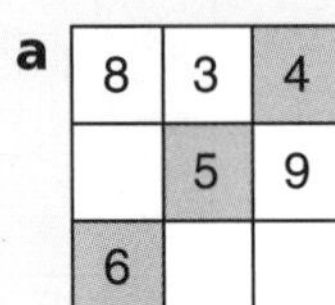

| 8 | 3 | 4 |
|---|---|---|
|  | 5 | 9 |
| 6 |  |  |

b

| 9 |  | 5 |
|---|---|---|
| 2 |  | 10 |
| 7 |  | 3 |

c

| 8 |  |  |
|---|---|---|
| 9 | 7 |  |
|  |  | 6 |

Add any line to find the sum.



 ISBN 978 0 6557 0886 5

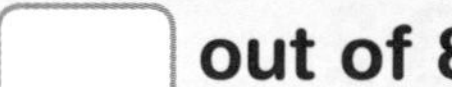

## 22:3 ☐ out of 8

1. Write an estimate to the nearest whole number in the next algorithm, then answer both questions.

   a
   ```
     7·108
   + 2·897   + ______
   ```
   b
   ```
     5·805
   + 3·868   + ______
   ```

2. I jumped 98 cm, 98 cm, 1 m, and 1 m 11 cm. What was the sum of the jumps? ______

3. The value of the 2 in:

   a 6·32 ______ b 70·25 ______

4. Find the size of the reflex angle. ______

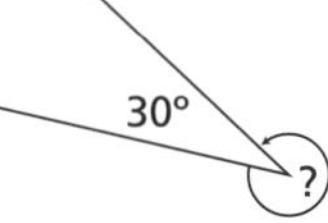


5. I poured six 0·7 L cups of water into the jug. How much did I pour into the jug? ______

6. Cara begins with $95.50. She buys:

   a What is Cara's total cost? ______

   b How much money is left? ______

   c If Cara buys this amount of milk and bread once a week, how much would she spend on this in 7 weeks? ______

7. How heavy are 8 bags of flour each weighing 4·75kg. ______

8. (28 − 14) + (89 − 67) ______

## 22:4 ☐ out of 7

**Extension**

1.

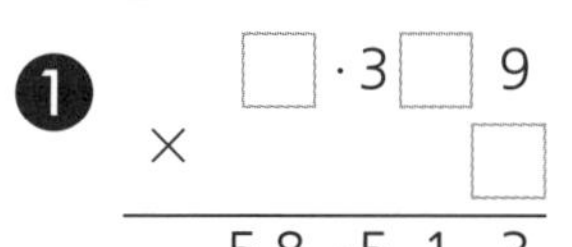


2.

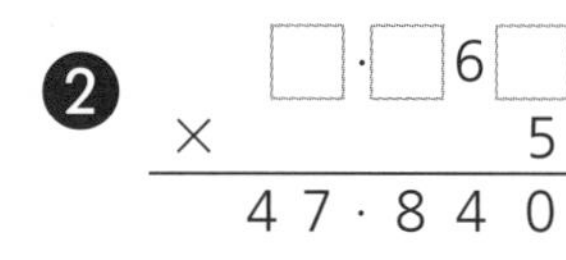


3. I spent 2 hours and 15 minutes on my laptop in the morning and again in the evening, 5 days a week. On the other 2 days, I spent 3 hours and 45 minutes each day. For how long did I use my laptop over a four week period? ______

4. A $1799 laptop was on sale for 25% off. How many could be bought with a budget of $20 000? ______

5. I have 5 shirts and 4 ties. In how many ways can I choose one of each? ______

6. 4 oranges cost $5.40 and a banana costs 40c more than an orange. David bought 5 oranges and 7 bananas. He paid and his change was $30. How much money did he start with? ______

7. Estimate the number that would be at point A on the number line. ______

   8966·5 —— A —— 8967·8

**Challenge**

*Write questions, using different operations, that are equal to:*

a 56 − 47

= ______

= ______

= ______

= ______

b 93 − 29

= ______

= ______

= ______

= ______

### Calculator crossnumber puzzle

**Across**

1 714 × 33

5 6463 − 5917

6 975 + 3618

8 7384 ÷ 568

9 $25^2$ + 240

10 $675^2$

**Down**

1 4148 ÷ 17

2 6000 − 807

3 625 ÷ 25

4 328 600 + 240 945

7 $72^2$

9 6000 − 5194

## 23:1 ☐ out of 22

1. 0·3 + 0·7 ______
2. 5 − 0·5 ______
3. $\frac{1}{4} + \frac{1}{4}$ ______
4. 63 + ______ = 100
5. $\begin{array}{r} 4{\cdot}678 \\ +\ 0{\cdot}856 \\ \hline \end{array}$
6. 756 − 136 ______
7. $\frac{1}{4}$ of 28 ______
8. 600 − 78 ______
9. 81 ÷ 9 × 9 ______
10. $\begin{array}{r} 8{\cdot}956\,\text{km} \\ -\ 0{\cdot}86\phantom{\,\text{km}} \\ \hline \end{array}$
11. $9\overline{)3875}$
12. $4\overline{)26606}$
13. Which is smallest, 7·77, 7·17 or 7·7? ______
14. 9 ÷ (45 − 42) + 34 ______
15. Write the improper fraction for:
    a $3\frac{4}{5}$ ______
    b $2\frac{1}{8}$ ______
16. Round 7 465 768 to the nearest million. ______
17. How many tens can be taken from:
    a 65 000 ______
    b 75 350 ______
18. True or false?
    a 576 + 765 = 765 − 576 ______
    b 78 × 34 = 34 × 78 ______
19. I gave 8 pencils to all my friends. If I gave out 72 pencils, how many friends do I have? ______
20. Change 560 mm to centimetres. ______
21. What number is 17 less than 100? ______
22. The value of 6 in 354·768. ______

## 23:2 ☐ out of 20

1. $\frac{16}{10} - \frac{2}{5}$ ______
2. $\frac{3}{4}$ of 24 ______
3. 67 × 57 ______
4. 99 × 34 ______
5. $\begin{array}{r} 7{\cdot}308\,\text{m} \\ \times\ \ \ \ \ \ \ 6\phantom{\,\text{m}} \\ \hline \end{array}$
6. 0·5 − 0·25 ______
7. 1345 + ______ = 1600
8. 1546 − ______ = 1256
9. 20% of $600. ______
10. $\begin{array}{r} 4{\cdot}726\,\text{cm} \\ \times\ \ \ \ \ \ \ 9\phantom{\,\text{cm}} \\ \hline \end{array}$
11. $10\overline{)96890}$
12. $6\overline{)6966}$

13. I throw one standard dice. Write, as a fraction, my chance of rolling a number less than 4. $\frac{\square}{\square}$
14. (3 + ______) × 7 = 35
15. 3 days from now it will be Saturday. What day was it 2 days ago? ______
16. $(5 \times 10^4) + (6 \times 10^3) + (4 \times 10^2) + (9 \times 10^1) + 9 =$ ______
17. Find the size of the reflex angle. ______

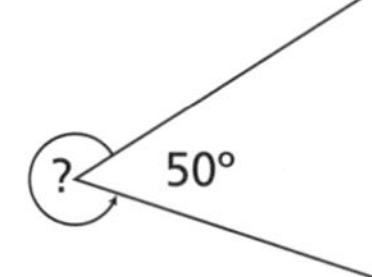


18. I printed a document 6 times. I used 9006 pieces of paper How many pages were in the document if I printed on only one side of the paper? ______

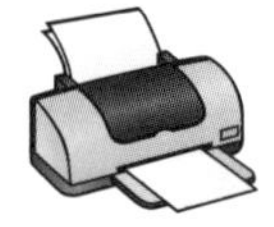

19. Change 3456 mm to centimetres. ______
20. I have $56 867 in my bank account. How many $10 notes could I withdraw? ______

## 23:3 ☐ out of 11

1. Write an estimate to the nearest whole number in the next algorithm, then answer both questions.

   a
   $$\begin{array}{r} 6{\cdot}387 \\ +\ 9{\cdot}576 \\ \hline \end{array} \quad + \_\_\_\_$$

   b
   $$\begin{array}{r} 8{\cdot}99 \\ +\ 4{\cdot}897 \\ \hline \end{array} \quad + \_\_\_\_$$

2. $10\overline{)25460}$

3. $3\overline{)9036}$

4. I throw one standard dice. Write, as a fraction, my chance of rolling a prime number. $\frac{\square}{\square}$

5. 8 liquid paper containers held 200 mL. What was the capacity of each container? ______

6. Eight people shared 1L of water. What is a fair share? ______

7. Change 3567 mm to centimetres. ______

8. 5400 people were placed in groups of 10 How many groups were formed? ______

9. 9 of my friends and I paid $4580 for band fees this term. If we each paid the same amount, how much did we each pay? ______

10. a 467 km = ______ m
    b 5763 g = ______ kg
    c 7·8 t = ______ kg
    d 4657 mm = ______ m
    e 8674 mL = ______ L
    f 6 hectares = ______ $m^2$
    g 78·4 km = ______ m

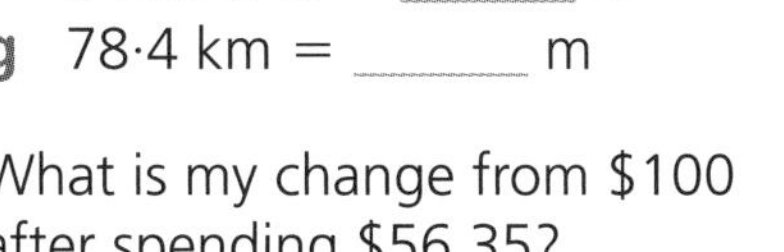

11. What is my change from $100 after spending $56.35? ______

## 23:4 ☐ out of 8

**Extension**

1. Circle the numbers that are divisible by both 4 and 5.

   610 305 1444 1600 420

2. The average of 13 numbers is 4·6. What is their sum? ______

3. It cost me $2304 to replace all the tyres on my car.
   a What was the cost of one tyre? ______
   b If the tyres were 20% off, how much would I save? ______

4. Train A is travelling at 38 km/h. Train B is travelling at 82 km/h. If they each travelled at this speed for an hour, how much further would Train B travel? ______

5. My heart beats 5040 times in an hour. How many times would it beat at this rate in 6 minutes? ______

6. 3 days from now it will be Tuesday. What day was it 30 days ago? ______

7. $65 \times 4 + 234 \div 9$ ______

8. To the total of 35, 68 and 35, add the product of 8 and 14. ______

**Challenge**

*Toss 2 coins 25 times and colour a box for each head or tail thrown.*

| | | | | | | | | | | | | | | | | | | | | | | | | | |
|---|---|---|---|---|---|---|---|---|---|---|---|---|---|---|---|---|---|---|---|---|---|---|---|---|---|
| Heads | | | | | | | | | | | | | | | | | | | | | | | | | |
| Tails | | | | | | | | | | | | | | | | | | | | | | | | | |

*What was the total of:*

a *heads?* ______ b *tails?* ______

*What was the percentage of:*

a *heads?* ______ b *tails?* ______

Complete this grid to write the distances in three ways.

| | | |
|---|---|---|
| 8·375 km | 8 km 375 m | 8375 m |
| | | 2914 m |
| 5·446 km | | |
| | 9 km 125 m | |
| | | 3546 m |
| 9·897 km | | |

## 24:1

 out of 17

1. 900 − 67 ______
2. ______ + 51 = 89
3. 81 ÷ 9 + 17 ______
4. 12 × 5 − 35 ______
5. $\begin{array}{r} 2{\cdot}613 \\ \times \quad\quad 2 \\ \hline \end{array}$
6. $\frac{8}{10} - \frac{1}{2}$ ______
7. 3·76 × 100 ______
8. $\frac{1}{4}$ of 32. ______
9. 78 − ______ = 34
10. $\begin{array}{r} 7{\cdot}024 \\ \times \quad\quad 3 \\ \hline \end{array}$
11. $10\overline{)7459{\cdot}0}$
12. $3\overline{)27{\cdot}09}$
13. **a** 456·43 ÷ 10 ______
    **b** 9·345 × 100 ______
    **c** 70·6384 ÷ 1000 ______
    **d** 46·5678 × 100 ______
14. How many tens in 34 567? ______
15. Measure the length of this bar.

    ______ mm or ______ cm
16. **a** 4·5 cm = ______ mm
    **b** 78 mm = ______ cm
    **c** 4300 m = ______ km
    **d** 6·8 km = ______ m
17. Which group of shapes below can be rearranged to make the flag on the right? ______

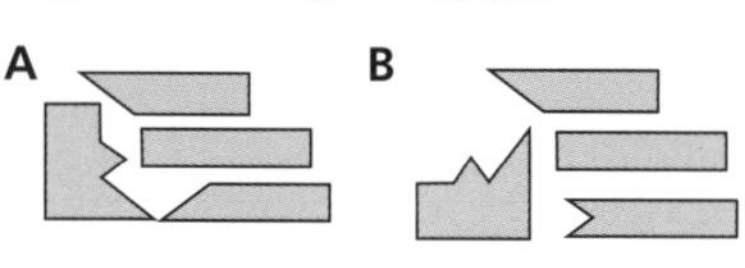


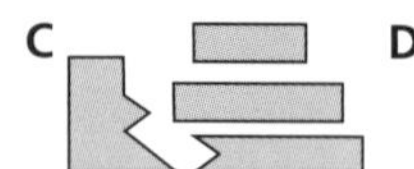


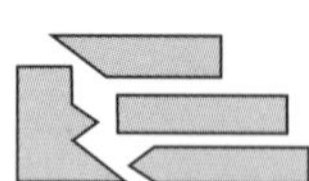

## 24:2

out of 18

1. $\frac{5}{6}$ of 54 ______
2. $8^2 - 48$ ______
3. $\frac{3}{4}$ of 100 ______
4. 7·896 ÷ 10 ______
5. $\begin{array}{r} 3{\cdot}496\,\text{m} \\ \times \quad\quad 7 \\ \hline \end{array}$
6. 0·6 + 1·4 ______
7. \$15 − \$3.67 ______
8. $\frac{18}{10} - \frac{8}{5}$ ______
9. 268 + ______ = 867
10. $\begin{array}{r} 5{\cdot}989\,\text{cm} \\ \times \quad\quad 8 \\ \hline \end{array}$
11. $10\overline{)375008}$
12. $7\overline{)89{\cdot}04}$
13. 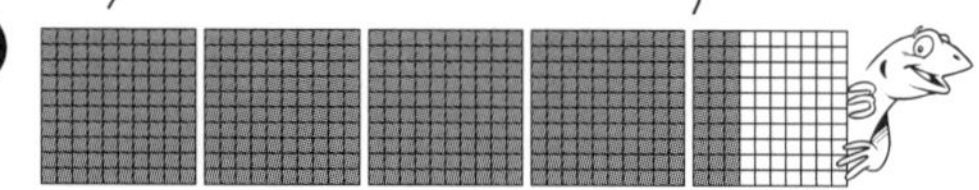

    Write this number as a:
    **a** mixed number ______
    **b** improper fraction ______
    **c** decimal ______
14. 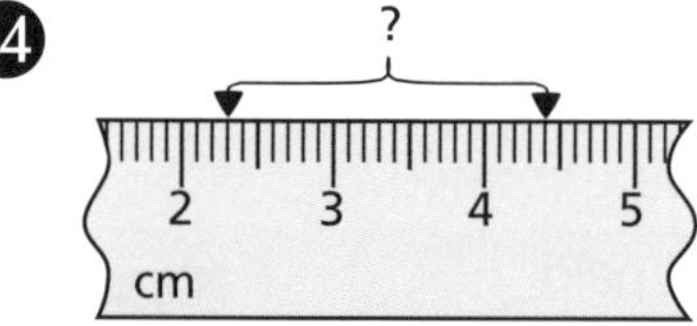

    Write the difference between the 2 marks on the ruler in:
    **a** centimetres ______
    **b** millimetres ______
15. Aimee ran an average of 4·5 km each day for 5 days. What was the total distance she ran? ______
16. **a** 876·45 ÷ 10 ______
    **b** 578·934 × 100 ______
17. **a** $\frac{1}{2}$ km = ______ m **b** $1\frac{3}{4}$ km = ______ m
18. What is my speed if I travel 178 km in 2 hours? ______

### Using a model

Alan, Rhonda, Rachel and Heather will sit on the four seats shown. Alan is in seat 3. Rhonda will sit between Rachel and Alan. Who will sit in seat number:

**a** 1? ______ **b** 2? ______

**c** 4? ______

## 24:3 ☐ out of 11

1.
$$\begin{array}{r} 5467884 \\ 208556 \\ 4678 \\ +\ 987967 \\ \hline \end{array}$$

2.
$$\begin{array}{r} 5{\cdot}989\,\text{cm} \\ \times\quad 5 \\ \hline \end{array}$$

3. $10\overline{)2658{\cdot}0}$

4. $9\overline{)26{\cdot}91}$

5. a 6247·21 ÷ 100 ______

   b 46·5678 × 1000 ______

6. What is the length of the black line in:

   a centimetres? ______

   b millimetres? ______

   0 cm 1 2 3 4

7. What is my speed if I travel 369 km in 3 hours? ______

8. I have walked 350 m of my 4·7 km journey. How much further have I to go? ______

9. a 9·7 km = ______ m

   b 367 mm = ______ cm

   c 8470 m = ______ km

   d 13·7 cm = ______ mm

   e $5\frac{4}{10}$ km = ______ m

10. The chance of Sophie scoring a goal at netball is about 1 in 3. About how many did she miss if she scored 7 goals? ______

11. 24 × 20 = (20 × ____) + (____ × ____)

    = ______ + ______

    = ______

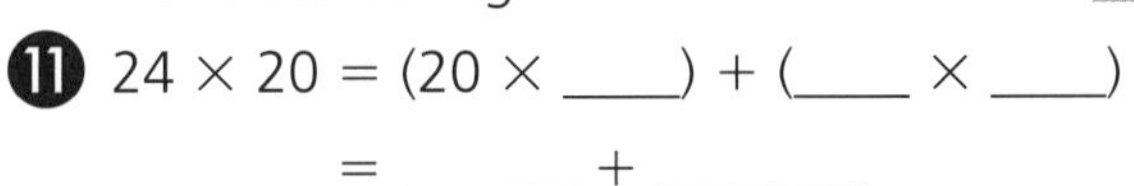

**Extension**

## 24:4 ☐ out of 6

1. How many millimetres in 6 m? ______

2. For our excursion, 3 buses took our group 90 km to our destination. Bus A had an average speed of 60 km/h. Bus B had an average speed of 52 km/h and Bus C had an average speed of 45 km/h. How long did the passengers of Bus A have to wait at the destination until Bus C arrived? ______

3. □ − △ + ▭ = 20
   If □ − △ = 13 and □ + ▭ = 25,
   find the value of △. ______

4. I am walking to my Nana's house, which is 2·5 km from my place. I will stay an hour and then walk back home. If I am 467 m from my house, how much further will I have to walk to get there and then back home again? ______

5. Sue has emus and horses. These have 40 legs altogether. What is the greatest number of horses possible? ______

6. What is the largest number of squares you can make using 22 unbroken paddle pop sticks? ______

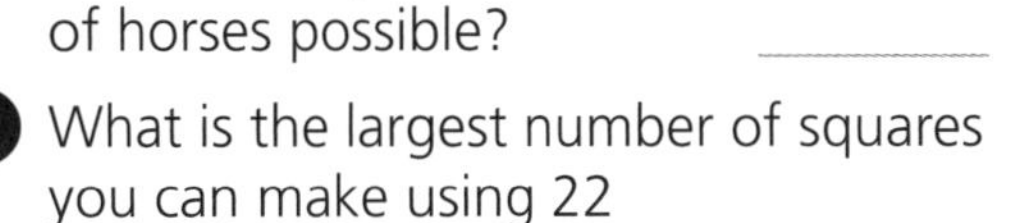

**Challenge**

Complete:

| | | | |
|---|---|---|---|
| 63·789 × 100 | ______ | 102·3 ÷ 10 | ______ |
| 49·028 ÷ 10 | ______ | 32·41 × 10 | ______ |
| 8·7034 × 1000 | ______ | 3409·8 ÷ 1000 | ______ |
| 9·416 × 100 | ______ | 429·37 ÷ 100 | ______ |
| 31·08 × 10 | ______ | 30·002 × 100 | ______ |
| 9002·3 ÷ 1000 | ______ | 5·1123 × 1000 | ______ |
| 8·325 × 100 | ______ | 9007·6 ÷ 1000 | ______ |
| 793·2 ÷ 10 | ______ | 2·31 × 10 | ______ |

Arkos wanted to find the thickness of a sheet of paper.
He found that the thickness of the pages in a 200-page book was 1 cm.

**a** How many sheets of paper make 200 pages? ______

**b** Complete this table.

| Thickness | 10 mm | 1 mm | 0·1 mm |
|---|---|---|---|
| Number of sheets | 100 | | |

**c** How thick would a book of 160 pages be? ______

## 25:1   ☐ out of 15

1. 80 − 59 ____
2. 800 − 59 ____
3. 0·7 + 1·3 ____
4. 7 × 8 − 30 ____
5. 365·46 − 95·48
6. $\frac{10}{12} - \frac{1}{6}$ ____
7. 36·7 × 10 km ____
8. 4°C less than 0 ____
9. 45 + ____ = 97
10. 746·24 + 36·79
11. Round the numbers in questions 5 and 10 to the nearest whole number to estimate and check your answers.
    a Round
    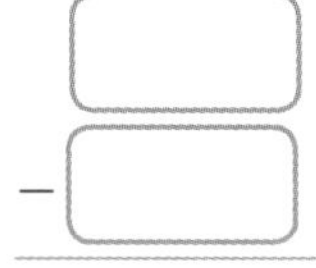
    b Round
    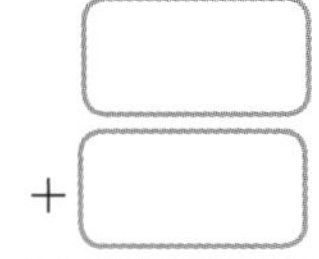
12. a 500 ÷ 50   (÷ 10) (÷ 10)   ____ ÷ ____ = ____
    b 4·9 ÷ 0·7   (× 10) (× 10)   ____ ÷ ____ = ____
13. The length of this pencil.
    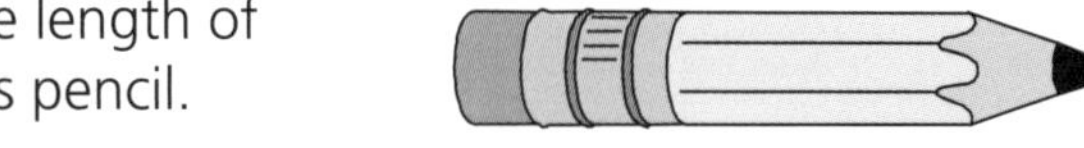
    a Estimate = ____ mm
    b Measure = ____ mm or ____ cm
14. a 6247·21 ÷ 10 ____
    b 46·5678 × 100 ____
15. a 59·9 cm = ____ mm
    b 8769 mm = ____ cm
    c 2·5 km = ____ m
    d $8\frac{2}{10}$ km = ____ m

## 25:2   ☐ out of 16

1. 81 ÷ 9 × 6 ____
2. $\frac{4}{10}$ of 50 L ____
3. 45·9 ÷ 10 ____
4. $8 - \frac{4}{5}$ ____
5. 263078 + 26478
6. $\frac{11}{12} - \frac{1}{2}$ ____
7. 62 × 74 ____
8. 54 ÷ ____ = 9
9. $7 − $3.45 ____
10. 297865 − 74986
11. Write 586 cm as metres using a decimal. ____
12. a 900 ÷ 30   (÷ 10) (÷ 10)   ____ ÷ ____ = ____
    b 3·6 ÷ 0·6   (× 10) (× 10)   ____ ÷ ____ = ____
13. Round to the nearest whole to estimate:
    a 67·5 − 28·9 ____
    b 35 ÷ 6·936 ____

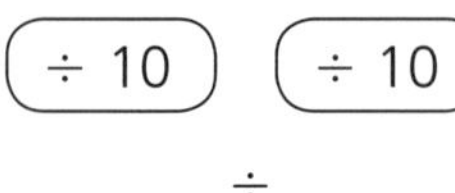

14. a 356·6 ÷ 100 ____
    b 948·23 × 1000 ____
15. Estimate, then measure the length of this line.

    a Estimate = ____ mm
    b Measure = ____ mm or ____ cm
16. a 56·9 m = ____ cm
    b 87 654 m = ____ km
    c 3756 mm = ____ cm
    d 27·9 cm = ____ cm ____ mm
    e 8·846 km = ____ m

### Estimation

Choose a measurement from the table as an estimate for the length of a:

A bed ____   B mouse ____
C bus ____   D paperclip ____
E shoe ____   F cricket bat ____
G car ____   H ant ____

| | |
|---|---|
| 2 cm | 10 m |
| 0·5 cm | 0·5 km |
| 2 m | 50 m |
| 2 km | 4 m |
| 100 km | 8 cm |
| 20 cm | 10 km |
| 1 mm | 75 cm |

## 25:3 [ ] out of 11

1.
```
  3768567
   106587
     2758
+  357003
```

2.
```
  $3567.68
− $ 426.03
```

3. Write 926 cm as metres using a decimal. ______

4. Each parking space in a car park is 230 cm wide. What is the width of 19 car spaces placed side by side? (Write your answer in metres.) ______

5. **a** 128 ÷ 16

______ ÷ ______

= ______

**b** 54 ÷ 18

÷ 2 ÷ 2

______ ÷ ______

= ______

6. Round to the nearest whole to estimate:

**a** 90·3 − 73·6 ______

**b** 81 ÷ 8·715 ______

7. Write in order, from smallest to largest:

$\frac{1}{2}$, $\frac{1}{12}$, $\frac{1}{3}$, $\frac{1}{6}$, $\frac{1}{4}$ ______

8. **a** 892·89 ÷ 10 ______

**b** 79·254 × 1000 ______

9. **a** 35·27 ÷ 10 ______

**b** 872·1 ÷ 100 ______

**c** 9138 ÷ 1000 ______

10. **a** 38·2 cm = ______ cm ______ mm

**b** 36·67 km = ______ m

11. (3 × 70) + (9 × 400) = ______

## 25:4 [ ] out of 3 — Extension

1. (45 × 67) + (47 × 89) × 3 ______

2. Sami is 137 cm tall. How many more millimetres does he need to grow to reach 170 cm? ______

3. Jedda needs to drop off parcels at **A**, **B**, **C** and **D**.

**a** Calculate the shortest route if he starts and ends at **E**. ______

**b** His average speed is 60 km/h. How long will this trip take, to the nearest minute. ______

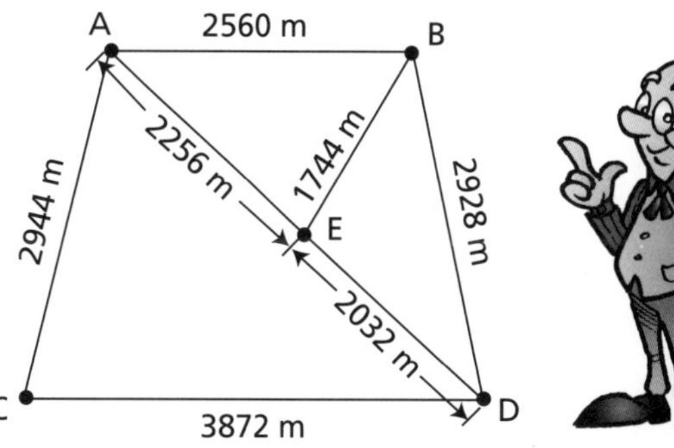


### Challenge

*Use a digital map (e.g. Whereis.com) to find the distance and travel time to visit landmarks from your location.*

**Example of factors**

Draw the rectangles that have an area of 6 units².

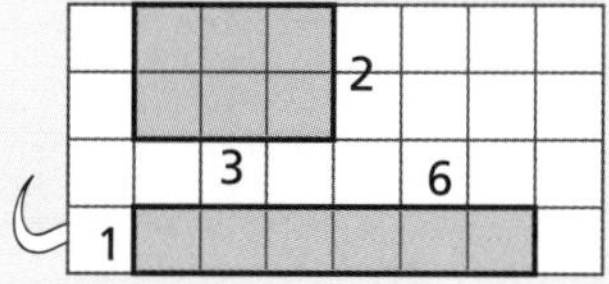

Draw rectangles of area 12 units².

List the factors. ______

## 26:1 ☐ out of 18

1. 456 + 143 ______
2. 748 − 234 ______
3. 7 × (3 + 7) ______
4. $5 - \frac{1}{2}$ ______
5. $6^2$ plus 30. ______
6. $\frac{1}{2}$ of 62. ______
7. 54 + ______ = 78
8. $\frac{7}{12} - \frac{2}{6}$ ______
9. $7\overline{)65{\cdot}94}$
10. $10\overline{)354657{\cdot}0}$
11. Estimate, then measure the length of this line.

    a Estimate = ______ mm

    b Measure = ______ mm or ______ cm
12. a 5·6 cm = ______ cm ______ mm

    b 78·72 km = ______ km ______ m
13. Round 3·876 to the nearest whole number ______
14. Colour $\frac{1}{3}$ of this shape red and $\frac{1}{4}$ of the shape blue. How much is coloured altogether?

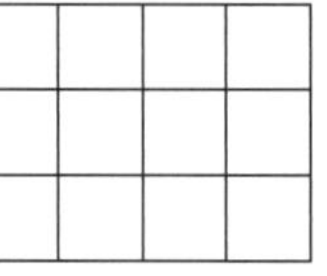

______

15. 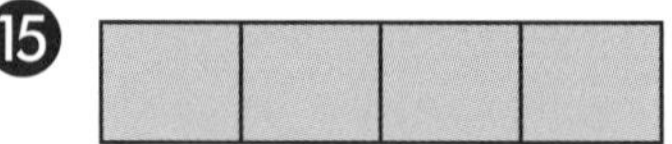

    a $\frac{1}{4} + \frac{2}{4} = \frac{☐}{☐}$   b $\frac{2}{4} + \frac{2}{4} = \frac{☐}{☐}$
16. a 546·21 ÷ 10 ______

    b 867·34 × 100 ______
17. How many tens can be taken from:

    a 573? ______   b 86 846? ______
18. (3 × 50) + (7 × 30) = ______

## 26:2 ☐ out of 19

1. 63 ÷ 9 × 7 ______
2. 3·7 L × 10 ______
3. $\frac{9}{12} - \frac{4}{6}$ ______
4. \$50 − \$12 ______
5. 78 − ______ = 27
6. 56 − 34 = ______ − 30
7. 81 + 73 = ______ + 70
8. $\frac{6}{7}$ of 49 mL ______
9. $5\overline{)91{\cdot}55}$
10. $10\overline{)71\ 089{\cdot}0}$
11. Write 375 cm as metres using a decimal. ______
12. My hand span is 18·4 cm long. My desk measures 4 hands spans plus 5·6 cm more. How long is my desk? ______
13. Divide 867·34 by ten. ______
14. 8·3 cm = ______ cm ______ mm
15. Estimate, then measure the length of this line.

    a Estimate = ______ mm

    b Measure = ______ mm or ______ cm
16. Naomi has 4 rulers of the same length. When placed in a line they have a length of 92 cm. How long is each ruler? ______
17. a 5·4 ÷ 0·9   b 560 ÷ 70

    (× 10) (× 10)   (÷ 10) (÷ 10)

    ______ ÷ ______ = ______   ______ ÷ ______ = ______
18. Round to the nearest whole to estimate:

    a 4·89 × 6·214 ______

    b 56 ÷ 6·894 ______

19. $\frac{3}{8} + \frac{1}{8} + \frac{2}{8} = \frac{☐}{☐}$

There are two eighths in 1 quarter.

Use the diagram to answer these questions.

a How many eights in one half? ______

b $1 - \frac{1}{4}$ ______   c $\frac{1}{2} + \frac{1}{8}$ ______   d $\frac{5}{8} - \frac{1}{8}$ ______

e $\frac{3}{4} - \frac{3}{8}$ ______   f $\frac{1}{2} + \frac{5}{8}$ ______   g $\frac{7}{8} - \frac{3}{4}$ ______

h $1\frac{7}{8} - \frac{1}{4}$ ______   i $2 - \frac{1}{8}$ ______   j $1\frac{1}{2} - \frac{5}{8}$ ______


 ISBN 978 0 6557 0886 5

## 26:3 ☐ out of 14

1. $\begin{array}{r} 5678{\cdot}57 \\ 573{\cdot}29 \\ +\ \ 36{\cdot}46 \\ \hline \end{array}$

2. $\begin{array}{r} 345{\cdot}67 \\ -\ 135{\cdot}79 \\ \hline \end{array}$

3. Emily is 65 mm taller than Sophie, and Sophie is 97 mm taller than Matthew. If Matthew is 1·450 m, how tall is Emily? ______
4. Round to the nearest whole to estimate:
   a 19·839 × 3·879 ______
   b 48·9 ÷ 6·85 ______
5. What is the chance as a percentage of throwing an even number using a dice made from this net? ______

| | 3 | |
|---|---|---|
| 5 | 1 | 5 |
| | 5 | |
| | 1 | |

6. (2 × 100) − (3 × 11 ) ______
7. Divide 367·7 by 100. ______
8. a $4 - \frac{1}{2}$ ______ b $8 - 1\frac{1}{2}$ ______
9. How many books with a spine width of 25 mm could I fit into my shelf that is 23 cm wide? ______
10. a 6297·3 × 10 ______
    b 834·97 × 100 ______
11. Melanie has 15 peaches, but $\frac{1}{5}$ of them were bad. How many were good? ______
12. 4·103 × 100 ______
13. < or >?
    a $\frac{1}{5}$ ☐ $\frac{3}{10}$ b $\frac{7}{10}$ ☐ $\frac{2}{5}$
14. $(7 \times 10^4) + (8 \times 10^3) + (9 \times 10^2) + (8 \times 10^1) + 4$
    = ______

## 26:4 Extension ☐ out of 6

1. I used 46 cm of ribbon to wrap a large present and 36 cm to wrap a small present. How much ribbon would I need to wrap 6 large presents and 7 small presents? ______
2. A can of soup costs $3.95. Alana bought 7 cans of soup and had $67.30 cash left over. How much did she start with? ______
3. Lin has 10 times as much money as Chris and he has $567 900 less than Harry. How much does Lin have if Harry has $876 580? ______
4. Write all possible pairs of counting numbers that make this sentence true.
   31 − 17= △ × □
   ______
5. What is the sum of the first 6 even counting numbers after 156. ______
6. Colour the shortest distance from **A** to **B**.

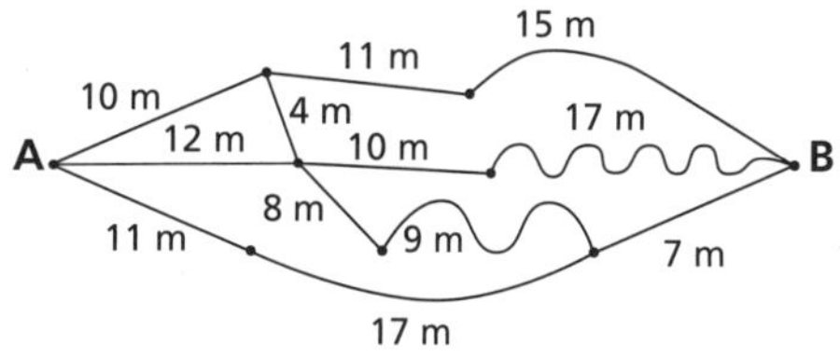

### Challenge

*Write facts about the number 13·452.*

______
______
______
______

### Concept

Use the diagram to answer these questions.

a $4 - \frac{1}{2} =$ ______ b $2 - \frac{1}{2} =$ ______

c $3 - \frac{3}{4} =$ ______ d $3 - \frac{1}{4} =$ ______

e $1 - \frac{1}{4} =$ ______ f $1 - \frac{3}{4} =$ ______

g $4 - \frac{2}{4} =$ ______ h $2 - \frac{3}{4} =$ ______

There are 16 quarters in 4 wholes.

## 27:1 □ out of 15

1. $5 \times$ ____ $= 35$
2. $\frac{1}{4}$ of 8. ____
3. $5 \times 60$ ____
4. $5{\cdot}6 \times 100$ ____
5. $(48 \div 8) + (36 \div 6)$ ____
6. $\frac{3}{4}$ of 36 m ____
7. $56 + 24 =$ ____ $+ 20$
8. $8 \times 900$ ____
9. $9\overline{)84{\cdot}69}$
10. $10\overline{)834569{\cdot}0}$
11. a $89{\cdot}3$ cm = ____ mm
    b 8345 m = ____ km
    c 2978 mm = ____ cm
    d $7{\cdot}2$ km = ____ m
    e $3\frac{3}{4}$ km = ____ m

See page 85 for help.

12. Find the perimeter of each shape.
    a (rectangle 24 cm × 9 cm) ____
    b (square 7 cm × 7 cm) ____
13. A B

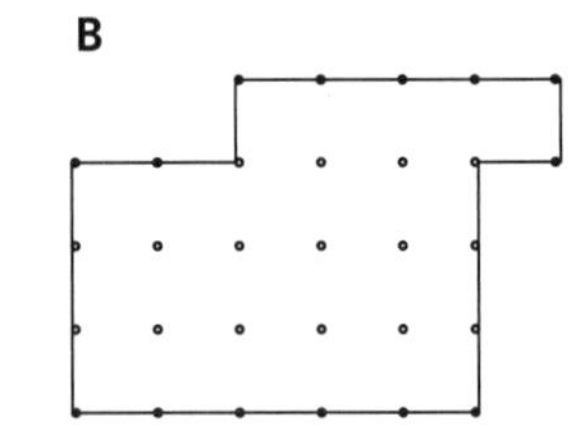

Which shape has the greater:
a area? ____ b perimeter? ____

14. a $\frac{4}{6} + \frac{1}{6}$ ____ b $\frac{10}{12} - \frac{5}{12}$ ____
15. a $15{\cdot}3 \div 10$ ____
    b $174{\cdot}762 \times 100$ ____
    c $8134{\cdot}6 \div 1000$ ____

## 27:2 □ out of 17

1. $435 + 645$ ____
2. $846 - 327$ ____
3. $6^2 - 2^2$ ____
4. $\frac{5}{6}$ of 54 L ____
5. $5 \times 6 -$ ____ $= 24$
6. $56 - 34 =$ ____ $- 30$
7. $56 + 34 =$ ____ $+ 30$
8. $8 + 20 \div 4$ ____
9. $5\overline{)85{\cdot}95}$
10. $10\overline{)290768{\cdot}0}$
11. $\frac{1}{5}$ of 20 kangaroos are male. How many are female? ____
12. $\frac{3}{4}$ of 1 litre. ____
13. a $\frac{13}{100} + \frac{4}{100}$ ____ b $\frac{18}{100} - \frac{9}{100}$ ____
    c $\frac{6}{100} + \frac{5}{100}$ ____ d $\frac{99}{100} - \frac{10}{100}$ ____
14. Find the area and perimeter of this shape.
    Area = ____
    Perimeter = ____

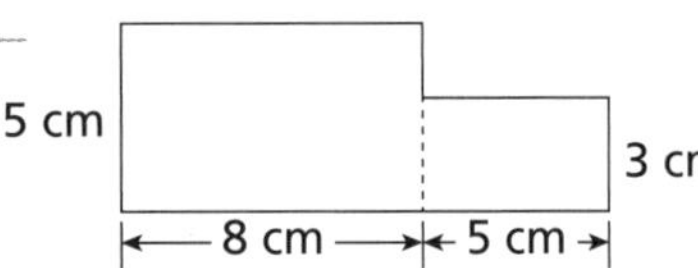


15. Write a fraction equal to:
    a $\frac{6}{10}$ □ b $\frac{1}{4}$ □ c $\frac{2}{6}$ □
16. a $\frac{3}{4}$ of 20 ____
    b $\frac{1}{5}$ of 20 ____
    c $\frac{2}{5}$ of 20 ____
    d $\frac{4}{5}$ of 20 ____
17. $0{\cdot}829$ km = ____ m

### Fractions to decimals using a calculator

Use a calculator to change these to decimals.

a $\frac{1}{2}$ ____ b $\frac{3}{5}$ ____ c $\frac{7}{20}$ ____
d $\frac{13}{25}$ ____ e $\frac{7}{8}$ ____ f $\frac{1}{16}$ ____
g $\frac{5}{8}$ ____ h $\frac{3}{16}$ ____ i $\frac{7}{25}$ ____
j $\frac{1}{32}$ ____ k $\frac{1}{64}$ ____ l $\frac{63}{64}$ ____

$\frac{3}{8} = 3 \div 8$ or $0{\cdot}375$

## 27:3   ☐ out of 9

**1**
```
     3546·78
      265·78
        3·56
 +    939·85
```

**2**
```
   530902
 − 253685
```

**3** Find the area and perimeter of this shape.

Area = ______

Perimeter = ______

(Shape: 12 m, 4 m, 5 m, 7 m)

**4** **a** $\frac{41}{100} + \frac{22}{100}$ ______   **b** $\frac{50}{100} - \frac{7}{100}$ ______

**c** $\frac{32}{100} - \frac{11}{100}$ ______   **d** $\frac{6}{100} + \frac{41}{100}$ ______

**5** Round each number to the nearest whole then estimate the answer.

**a** 3·4 × 7·2 ______

**b** 9·3 × 6·9 ______

**c** 26·8 − 13·1 ______

**6** Write all the factors of 42.

______

**7** Write a fraction equal to:

**a** $\frac{2}{5}$ ☐   **b** $\frac{4}{5}$ ☐   **c** $\frac{1}{5}$ ☐

**8** $\frac{1}{4}$ of 20 oranges. ______

**9** Asha spent $100 on party decorations. Tye only spent two tenths of what Asha spent. How much did Tye spend?

______

## 27:4   Extension   ☐ out of 5

**1** The area of:

**a** the shaded part? ______

**b** the part not shaded? ______

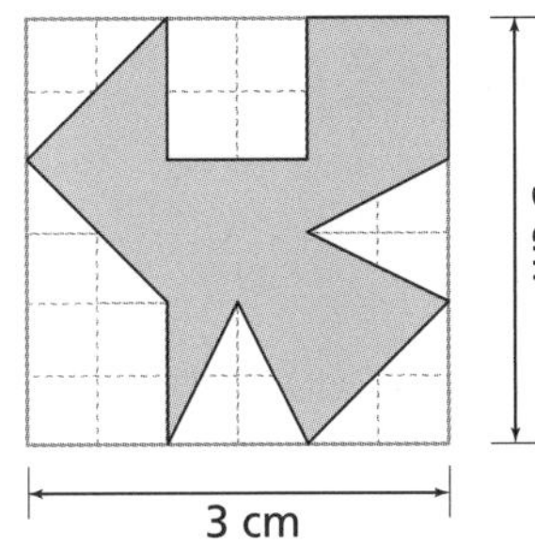


**2** A bird's head is $\frac{1}{3}$ as long as its body and $\frac{1}{4}$ as long as its tail. The bird is 48 cm long. How long is its head? ______

**3** In a test, Luis scored $\frac{3}{4}$ of the possible marks, while Peter scored $\frac{4}{5}$ of the possible marks. Which of the two boys scored the higher mark? ______

**4** Jodie ordered $\frac{1}{4}$ of a pizza for each of her 14 friends. How many pizzas did she order? ______

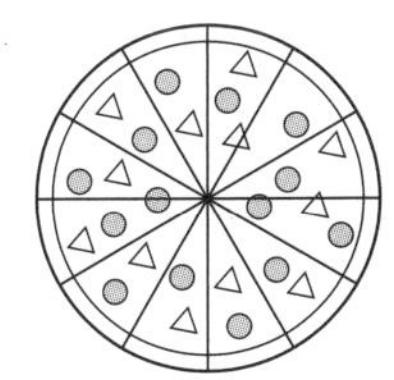

**5** **a** 30 squared ______

**b** 300 squared. ______

**Challenge**

*Complete this fraction wall.*
*Record equivalent fractions below.*

<table>
<tr><td colspan="4">$\frac{1}{2}$</td><td colspan="4">$\frac{1}{2}$</td></tr>
<tr><td colspan="2">$\frac{1}{4}$</td><td colspan="2"></td><td colspan="2"></td><td colspan="2"></td></tr>
<tr><td>$\frac{1}{8}$</td><td></td><td></td><td></td><td></td><td></td><td></td><td></td></tr>
</table>

Note: $\frac{2}{4} = \frac{4}{8}$

× Tables

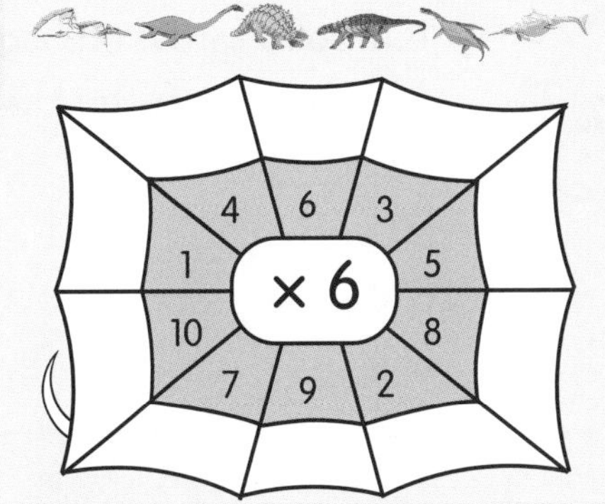

## 28:1 ☐ out of 16

1. $5{\cdot}6 \times 10$ ______
2. $746 - 235$ ______
3. $900 - 73$ ______
4. $5 \times 700$ ______
5. $\frac{7}{12} - \frac{1}{6}$ ______
6. $45 +$ ______ $= 100$
7. $40 \times 8$ ______
8. $\frac{1}{4}$ of 24 ______
9. $8\overline{)354672}$
10. $5\overline{)73{\cdot}65}$
11. What is the perimeter of this rectangle? ______

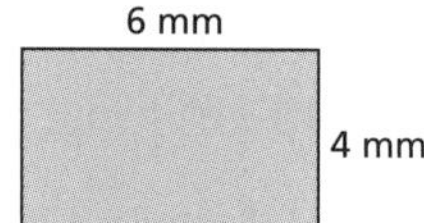


12. Write a fraction equal to:

a $\frac{8}{10}$ ☐ b $\frac{4}{10}$ ☐ c $\frac{6}{10}$ ☐

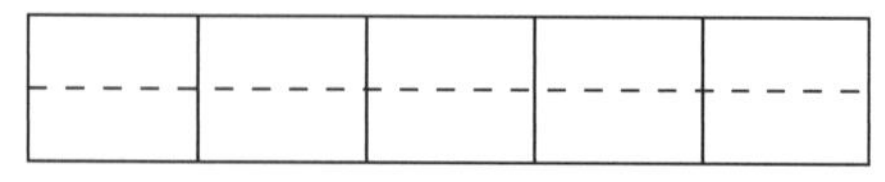

13.
a $6{\cdot}9$ cm = ______ mm
b 2947 m = ______ km
c 5456 mm = ______ cm
d $3{\cdot}7$ km = ______ m
e $1\frac{4}{10}$ km = ______ m

14. Kim's bus came at 7:50 am. She arrived at the station at 8:20 am. How long did the journey take? ______

15. Write 4:28 pm in 24-hour time. ______
16. Hours from 10 am to:

a 3 pm ______ b midnight ______
c 1 am ______ d 7:00 pm ______

## 28:2 ☐ out of 18

1. $526 - 313$ ______
2. Triple 57. ______
3. $6 + 56 \div 7$ ______
4. $1\frac{9}{10} - \frac{7}{5}$ ______
5. $5 + 3 \times 5$ ______
6. $167 +$ ______ $= 219$
7. $6\text{ L} \times 800$ ______
8. $\frac{4}{6}$ of 18 ______
9. $7\overline{)398062}$
10. $4\overline{)37{\cdot}76}$
11. a

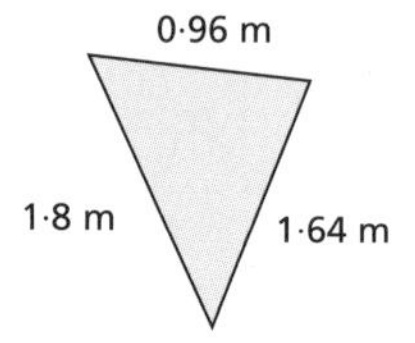


Perimeter = ______

b

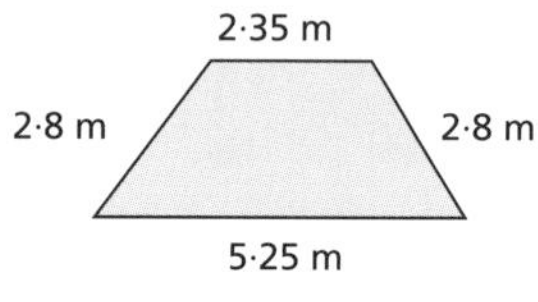


Perimeter = ______

12. a $\frac{1}{5} + \frac{3}{5} =$ ______ b $\frac{5}{8} - \frac{1}{8} =$ ______
13. 4 take away $1\frac{1}{4}$ ______
14. Which fraction is bigger:

a $\frac{3}{4}$ or $\frac{1}{3}$? ______ b $\frac{3}{4}$ or $\frac{2}{3}$? ______

15. a $\frac{4}{10}\frac{(\div 2)}{(\div 2)}$ ☐ b $\frac{2}{6}\frac{(\times 3)}{(\times 3)}$ ☐
16. What time has elapsed between 5 pm Tuesday and 2 am the following day? ______
17. $\frac{2}{3} - \frac{1}{6} = \frac{☐}{☐} - \frac{☐}{☐} = \frac{☐}{☐}$

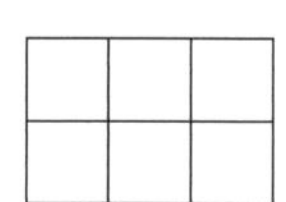

18. Use am or pm to write the time 1 hour after:

a 15:35 ______
b 03:09 ______

Turn to ID card A on page 6.
Give the answers for these numbers.

(1) ______ (2) ______
(3) ______ (4) ______
(24) ______ (25) ______
(26) ______ (27) ______
(28) ______ : ______ ______ (29) ______ : ______

The metric system is based on the number 10.

 ISBN 978 0 6557 0886 5

## 28:3 ☐ out of 10

**1**

$$\begin{array}{r} 5476899 \\ 246567 \\ 97 \\ +\ \ 48670 \\ \hline \end{array}$$

**2**

$$\begin{array}{r} 286908 \\ -\ 175679 \\ \hline \end{array}$$

**3** What fraction of $1 is 30c? ______

**4** Write all the factors of 32. ______

**5** The first 10 multiples of 9 are:

______

**6** **a** 5 squared ______ **b** 7 squared ______

**7** **a** $\frac{7}{8} - \frac{2}{8}$ ______ **b** $\frac{3}{6} + \frac{2}{6}$ ______

**c** $\frac{3}{10} + \frac{4}{10}$ ______ **d** $\frac{7}{12} - \frac{2}{12}$ ______

**8** Write a fraction equal to:

**a** $\frac{1}{6}$ ☐ **b** $\frac{2}{8}$ ☐ **c** $\frac{4}{10}$ ☐

**d** $\frac{5}{5}$ ☐ **e** $\frac{1}{3}$ ☐ **f** $\frac{4}{5}$ ☐

**9** The value of 8 in 34·798 ______

**10** How long does it take to travel from:

**a** May St to Joy Ave? ______

**b** Red St to Rae St? ______

**c** May St to Rae St? ______

| Bus timetable | |
|---|---|
| May St | 08:54 |
| Joy Ave | 09:12 |
| Red St | 09:37 |
| Fig St | 09:48 |
| Rae St | 10:04 |

## 28:4 ☐ out of 6

**Extension**

**1** In 1984 my age was half my father's age. 23 years later my age was 21 years younger than my father. If my father was 65 at this time, in which year was:

**a** I born? ______

**b** my father born? ______

**2** Find the area and perimeter of this shape.

Area = ______

Perimeter = ______

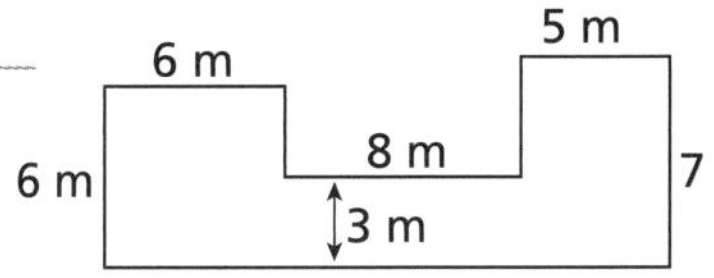


**3** $\frac{3}{4}$ of 8 L ______

**4** $\frac{7}{10} + \frac{1}{5} = \frac{\square}{20} + \frac{\triangle}{20}$ $\square =$ ______, $\triangle =$ ______

**5** Linda weighs 57·45 kg. She was holding $8\frac{1}{2}$ kg of groceries when she stepped onto the scales.

**a** What was the total weight? ______

**b** How much less than 100 kg is the total weight? ______

**6** **a** (265 − 49) + 63 = ______

**b** (176 − 128) ÷ 4 = ______

**Challenge**

*Draw and describe the features of a parallelogram.*

______

______

______

______

**a** To get to camp, Year 6 used 8-seater vans and 5-seater cars. There were 10 vehicles used. If there were 59 seats, how many vans were used? ______

**b** In the paddock, kangaroos and cows grazed together. The 31 animals had 100 legs altogether. How many of each type of animal were in the paddock?

kangaroos: ______ cows: ______

## 29:1 — ☐ out of 18

1. 100 − 37 ______
2. 7 × 80 ______
3. 4·5 × 10 ______
4. 5 × 3 + 7 ______
5. Factors of 17 ______
6. $\frac{1}{4}$ of 16 ______
7. 35 + ______ = 80
8. $\frac{5}{8} - \frac{1}{2}$ ______
9. $3\overline{)857032}$
10. $7\overline{)85{\cdot}75}$
11. Write a fraction equal to:
    a $\frac{2}{5}$ ☐  b $\frac{1}{2}$ ☐  c $\frac{3}{4}$ ☐
12. 15 birds are in a flock. One third fly away. How many are left? ______

13. $\frac{3}{4}$ plus $\frac{3}{4}$. ☐

14. What is the value of the bold digits?
    a 7**4**1 865 ______
    b 989 3**5**0 ______
    c 354·56**8** ______

15. How many tens can be taken from:
    a 356 798? ______
    b 8 745 325? ______
16. a $\frac{1}{4}$ plus $\frac{1}{4}$. ______
    b $\frac{1}{4}$ plus $\frac{1}{2}$ ______

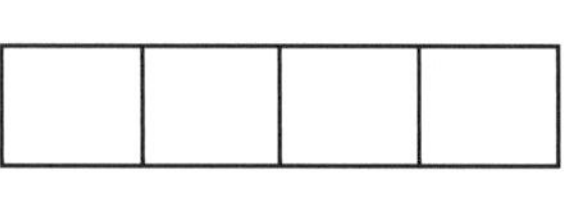

17. If a train leaves at 11:38 am and arrives at 12:12 pm, how long does the journey take? ______

18. What is 20% of 60? ______

## 29:2 — ☐ out of 21

1. 56 × 3 ______
2. $\frac{3}{4}$ of 36 L ______
3. 64 ÷ 8 + 9 ______
4. 85 − 43 − 8 ______
5. $\frac{4}{12} + \frac{3}{6}$ ______
6. 256 − 34 = ______ − 30
7. 7 × 700 ______
8. 25% of $24 ______
9. $7\overline{)285678}$
10. $8\overline{)25{\cdot}52}$
11. < or >? a $\frac{1}{2}$ ☐ $\frac{4}{6}$  b $\frac{3}{4}$ ☐ $\frac{3}{8}$

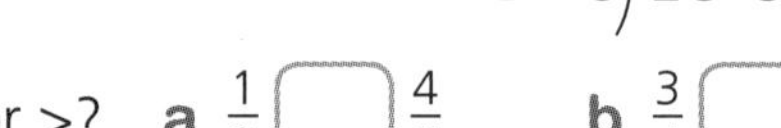

12. What is 30% of $50? ______
13. How many weeks in half a year? ______
14. Use 24-hour time to write 5:26 pm. ______
15. Show the digital and analog time for 19 minutes after 07:53.
    a

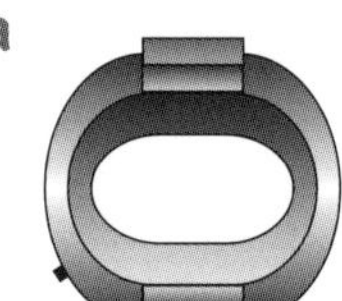

    b

16. $\frac{3}{10} + \frac{1}{2} = \frac{\square}{\square} + \frac{\square}{\square} = \frac{\square}{\square}$
17. Use am or pm to write the time 1 hour after:
    a 18:27 ______
    b 03:09 ______
18. Find the perimeter of a square with side lengths of 56·4 cm. ______
19. How many hours in 2 days? ______
20. Find the area of a square with side lengths of 9 m. ______
21. (56 + 35) − 5 × 5 ______

**Average speed = distance travelled for each unit of time**

Find the average speed if:

a 200 metres is covered in 10 seconds. ______ per second

b 144 kilometres is covered in 3 hours. ______ per hour

c 980 kilometres is covered in 4 days. ______ per day

d 60 millimetres is covered in 5 seconds. ______ per second

 • *AUSTRALIAN SIGNPOST MATHS 6 MENTALS* • ISBN 978 0 6557 0886 5

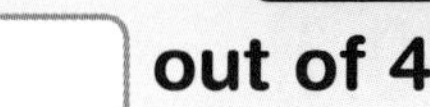

## 29:3 ☐ out of 11

1 
$$\begin{array}{r} 9456499 \\ 189567 \\ 2989 \\ +\ 234000 \\ \hline \end{array}$$

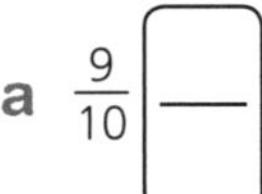

2 
$$\begin{array}{r} 703000 \\ -\ 296849 \\ \hline \end{array}$$

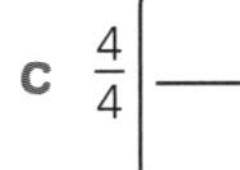

3 Use 24-hour time to write 5:26 pm. ______

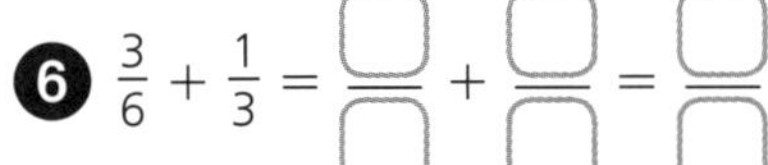
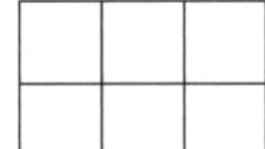

4 Write a fraction equal to:

a $\frac{9}{10}$ $\frac{\square}{\square}$ b $\frac{1}{3}$ $\frac{\square}{\square}$ c $\frac{4}{4}$ $\frac{\square}{\square}$

5 Use am or pm to write the time $4\frac{1}{2}$ hours after:

a 21:30 ______ b 02:09 ______

6 $\frac{3}{6} + \frac{1}{3} = \frac{\square}{\square} + \frac{\square}{\square} = \frac{\square}{\square}$

7 Show the digital and analog time for 2 hours and 35 minutes after 14:36.

a

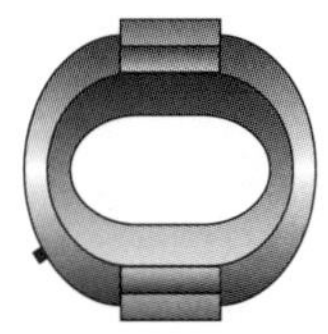

b

8 a Name this shape. ______

b Is this shape a quadrilateral? ______

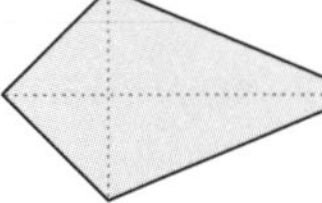

9 What is 40% of $70? ______

10

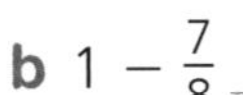

a $\frac{1}{2} + \frac{3}{8}$ ______ b $1 - \frac{7}{8}$ ______

11 Calculate the:

a area ______

b perimeter ______

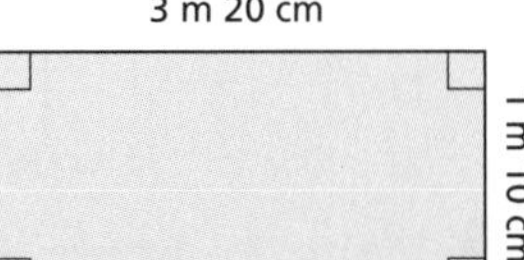


## 29:4 ☐ out of 4

Extension

1 One fifth of a whole is 9.
Write the value of the shaded part.

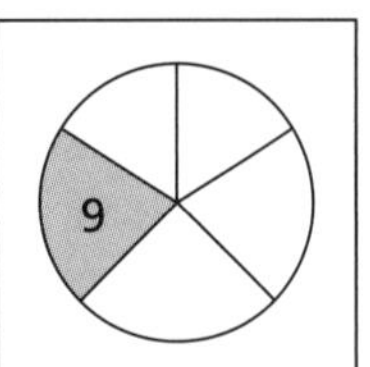


a

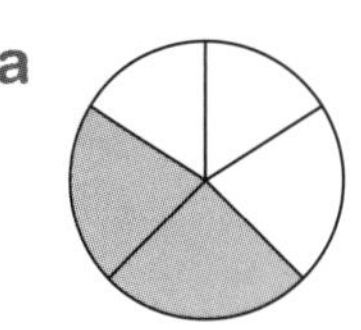

b

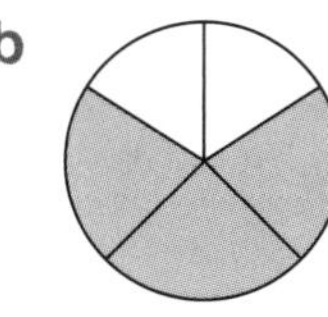

______ ______ ______

2 30% of the townspeople were men, 27% were women and 51% were male.
If there were 4000 people in total, how many were:

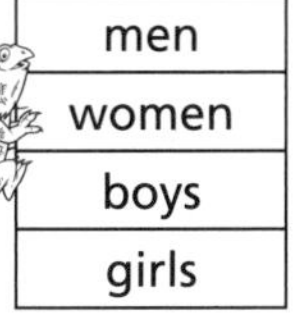

| men |
|---|
| women |
| boys |
| girls |

a male? ______ b boys? ______

c women? ______ d girls? ______

3 Compare these two shapes. What is the difference between the:

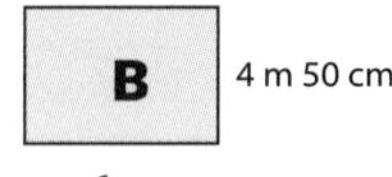


a perimeter of each? ______

b area of each? ______

4 These are finishing times for a race. Who came 2nd?

______

| Ty | 21:54·69 |
|---|---|
| Kim | 21:53·45 |
| Ron | 22:01·56 |
| Sue | 21:54·71 |

**Challenge**

*Write facts about the number* 356 467 453.

______

______

______

**Use the clues to fill in the tables.**

Put a tick for YES ☑
Put a cross for NO ☒

| | Drama | Gym | Tennis | Monday | Tuesday | Friday |
|---|---|---|---|---|---|---|
| **Heather** | | | | | | |
| **Naomi** | | | | | | |
| **Luke** | | | | | | |

Heather, Naomi and Luke are in different clubs. One of the girls goes to gym on Monday. Naomi goes to her club on Tuesday. Luke does not do drama. His club meets on Friday.

| ANSWERS | Club | Day |
|---|---|---|
| **Heather** | | |
| **Naomi** | | |
| **Luke** | | |

 • *AUSTRALIAN SIGNPOST MATHS 6 MENTALS* • ISBN 978 0 6557 0886 5 

## 30:1    ☐ out of 16

1. 534 + 253 ______
2. 867 – 253 ______
3. Triple 25. ______
4. $\frac{5}{6} - \frac{5}{12}$ ______
5. 20% of 100 ______
6. $\frac{3}{10} + \frac{4}{10}$ ______
7. 56 ÷ 7 + 13 ______
8. 100 × 6·5 cm ______
9. **a** $\frac{5}{7} = \frac{\square}{28}$, $\square$ = ______

   **b** $\frac{2}{7} = \frac{\square}{28}$, $\square$ = ______

   **c** $\frac{1}{4} = \frac{\square}{28}$, $\square$ = ______

   **d** $\frac{3}{4} = \frac{\square}{28}$, $\square$ = ______

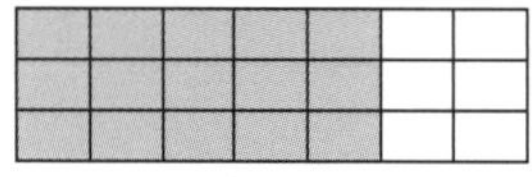

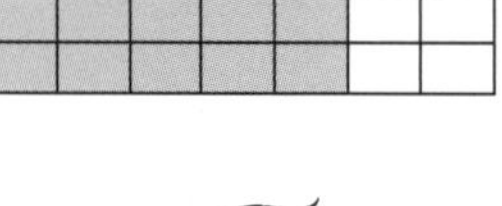

10. Use am or pm to write the time 3 hours after:

    **a** 23:24 ______    **b** 09:56 ______

11. A rectangle measures 7 cm by 8 cm. What is its:

    **a** perimeter? ______

    **b** area? ______

12. 

    For this cricket pitch find the:

    **a** area ______    **b** perimeter ______

13. Write this as 24-hour time.

______

14. What is 40% of $70? ______
15. (16 – 7) × (5 + 3) ______
16. **a** 29·6 cm = ______ mm

    **b** 2980 m = ______ km

## 30:2    ☐ out of 17

1. 81 ÷ 9 + 8 ______
2. 351 – 247 ______
3. 6 × 90 L ______
4. $\frac{2}{6} + \frac{1}{2}$ ______
5. $$\begin{array}{r} 456783 \\ +\ \ 96783 \\ \hline \end{array}$$
6. 25% of 50 ______
7. 72 ÷ 9 × 5 ______
8. 18 − 12 ÷ 4 ______
9. 76 + ______ = 116
10. $$\begin{array}{r} 935734 \\ -\ \ 94759 \\ \hline \end{array}$$
11. Write as 24-hour time:

    **a** 8:50 am ______    **b** 8:50 pm ______

12. $\frac{4}{5} - \frac{2}{10} = \frac{\square}{\square} - \frac{\square}{\square} = \frac{\square}{\square}$

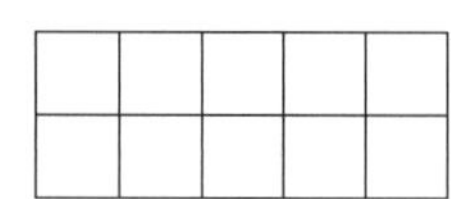

13. Our school has 10 soccer balls. 6 of them were used today. What percentage were not used? ______
14. Calculate the:

    **a** perimeter? ______

    **b** area? ______

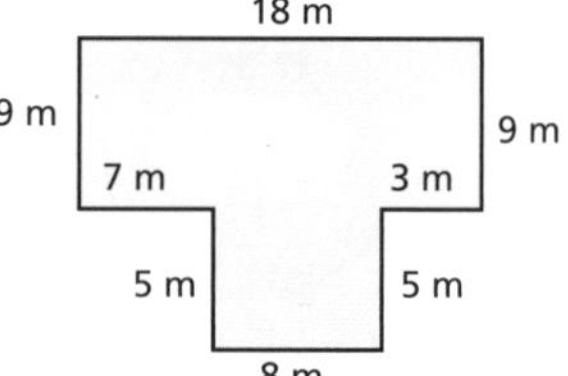


15. How many minutes in:

    **a** $1\frac{1}{2}$ hours? ______

    **b** $2\frac{1}{4}$ hours? ______

16. Find the lowest common denominator, then find:

    **a** $\frac{1}{2} + \frac{1}{3} = \frac{\square}{\square} + \frac{\square}{\square} = \frac{\square}{\square}$

    **b** $\frac{3}{4} + \frac{1}{5} = \frac{\square}{\square} + \frac{\square}{\square} = \frac{\square}{\square}$

17. Write $7\frac{1}{4}$ hours as a decimal. ______

1 How many triangles are in this grid that have a side length of:

**a** 1 unit? ______    **b** 2 units? ______    **c** 4 units? ______

2 What is the greatest number of small triangles that can be shaded if no two shaded triangles share:

**a** a side? ______    **b** a corner? ______

© PEARSON AUSTRALIA 2024 • *AUSTRALIAN SIGNPOST MATHS 6 MENTALS* • ISBN 978 0 6557 0886 5

## 30:3 ☐ out of 12

**1**
```
  4657809
   300800
     6859
+  269895
```

**2**
```
  306003
- 230035
```

**3** What is 5% of $20? ______

**4** Find the perimeter of a square with side lengths of 89·3 m. ______

**5** < or >?

a $\frac{1}{10}$ ☐ $\frac{4}{5}$  b $\frac{5}{6}$ ☐ $\frac{2}{3}$

**6** Aaron mowed the lawn from 3:39 pm till 5:06 pm. How long did he take?

______

**7** Find the area of a square with side lengths of 34 m. ______

**8** 50 + 20 ÷ (8 ÷ 2) − 24 ______

**9** Find the lowest common denominator, then find:

a $\frac{3}{5} - \frac{1}{2} = \frac{\square}{\square} - \frac{\square}{\square} = \frac{\square}{\square}$

b $\frac{3}{4} - \frac{1}{3} = \frac{\square}{\square} - \frac{\square}{\square} = \frac{\square}{\square}$

**10** Number line: C, −0·15, −0·1, A, 0, B, 0·1

What number would be at:

a **A**? ____  b **B**? ____  c **C**? ____

**11** Use am or pm to write the time 3 hours after:

a 13:51 ______  b 04:36 ______

**12** a 3·4 cm = ______ mm

b 8937 m = ______ km

c 845 mm = ______ cm

d 2·8 km = ______ m

## 30:4 Extension ☐ out of 6

**1** Ethan ate $\frac{1}{3}$ of the orange and Lucas ate $\frac{1}{4}$. What fraction of the orange was left? ______

**2** Sam raised $1657 more than Harry, and Riley raised half as much as Harry. What did they raise in total if Harry raised $658? ______

**3** Oliver had green, blue, white and red balls. $\frac{3}{12}$ were green, $\frac{1}{4}$ were blue, $\frac{1}{3}$ were white, and 4 were red.
How many were green or blue? ______

**4** 10% reduction was given on the price of a TV marked at $890. How much was paid? ______

**5** Guess my number.
76 less than a third of my number is divided by 5 to get 34. My number is ______.

**6** For this shape find the:

a perimeter? ______

b area? ______

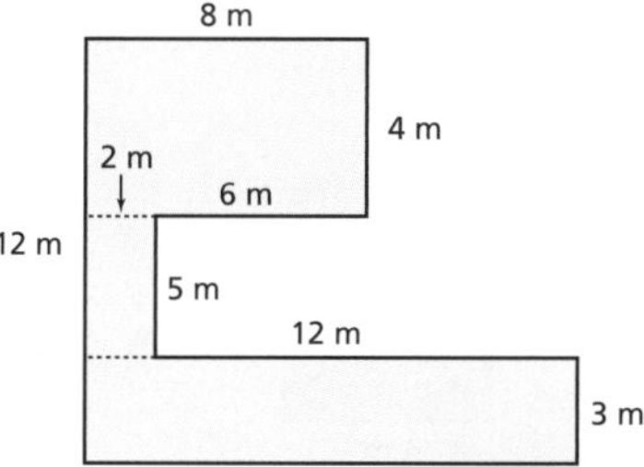


**Challenge**

*Make up questions like Question 5 above, to challenge your friends. Write the answers somewhere else in your book.*
*Guess my number:*

______

______

______

______

### Codes

Concept

Use the code to write 'true' or 'false' for each code statement.

| A | B | C | D | E | F | G | H | I | J | K | L | M | N | O | P | R | S | T | U | V | W | Y |
|---|---|---|---|---|---|---|---|---|---|---|---|---|---|---|---|---|---|---|---|---|---|---|
| 1 | * | 9 | & | 4 | ? | 0 | / | 2 | ( | + | 7 | < | 5 | # | ) | 3 | $ | = | 8 | @ | > | 6 |

**a** 1 7#50 (#83546 $=13=$ >2=/ (8$= #54 $=4). ______

**b** $8994$$ &4)45&$ 7130476 #5 /13& >#3+. ______

**c** <156 34942@4 1&@294, =/4 >2$4 )3#?2= ?3#< 2=. ______

## 31:1 ☐ out of 15

1. $6 \times$ ____ $= 36$
2. 50% of 24 ____
3. $7 \times 4 - 25$ ____
4. $3 \times 8 + 6$ ____
5. $\begin{array}{r} 4675{\cdot}2 \\ +\ 364{\cdot}7 \\ \hline \end{array}$
6. $\frac{1}{4} + \frac{3}{8}$ ____
7. $156 - 59$ ____
8. $\frac{1}{4}$ of 32 ____
9. $34 + 39 =$ ____ $+ 40$
10. $\begin{array}{r} 365{\cdot}18 \\ -\ 33{\cdot}24 \\ \hline \end{array}$
11. What is the area of a rectangular dance floor that has a length of 9 m and a width of 6 m?

____

12. Which of these is the net of an open cube?

____

13. Find the area of:

____ ____

14. Find the lowest common denominator, then find:

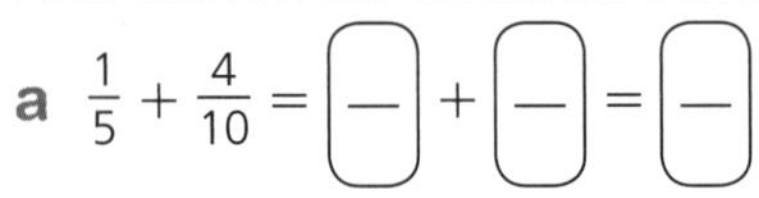
a $\frac{1}{5} + \frac{4}{10} = \frac{\square}{\square} + \frac{\square}{\square} = \frac{\square}{\square}$

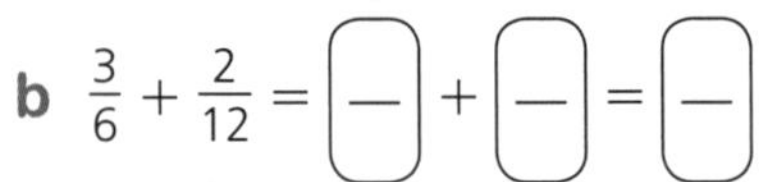
b $\frac{3}{6} + \frac{2}{12} = \frac{\square}{\square} + \frac{\square}{\square} = \frac{\square}{\square}$

15. Find the equal measures: (See page 85.)
   a 3 L = ____ mL
   b 9 L 261 mL = ____ mL
   c 2·734 L = ____ mL

## 31:2 ☐ out of 19

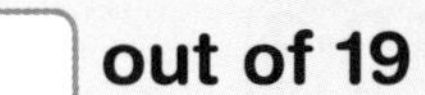

1. 20% of 50 ____
2. $\frac{1}{3}$ of 12 ____
3. $35 \times 20$ ____
4. $0{\cdot}6 \times 10$ ____
5. $5 \times 50$ ____
6. $42 \div 2 - 8$ ____
7. $58 + 147 =$ ____ $+ 150$
8. $271 - 38 =$ ____ $- 40$
9. $5\overline{)356455}$
10. $6\overline{)73{\cdot}38}$
11. What is the greatest number of shape A that could be cut out from shape B?

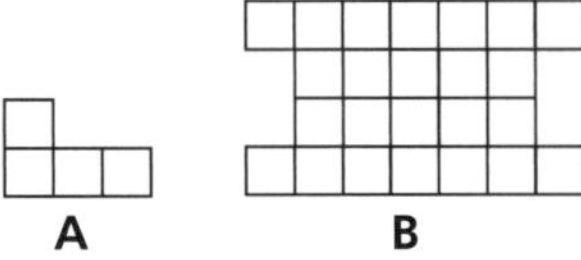


____

12. $2 \times 25 - 30 - 5 =$ ____
13. Find the lowest common denominator, then find:

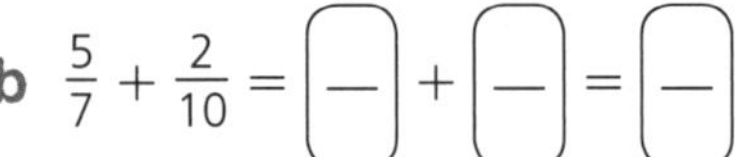
a $\frac{8}{10} - \frac{1}{2} = \frac{\square}{\square} - \frac{\square}{\square} = \frac{\square}{\square}$

b $\frac{5}{7} + \frac{2}{10} = \frac{\square}{\square} + \frac{\square}{\square} = \frac{\square}{\square}$

14. Use decimals to write these as litres.
   a 386 mL = ____ L
   b 946 mL = ____ L
15. a $7 + 3\frac{1}{2}$ ____ b $6 + 4\frac{1}{2}$ ____
16. What is 25% of $40? ____
17. What is the cost of 13 balls if they are $8 each? ____
18. If the time is 8:39 pm on December 31st, how long will it be until the new year? ____
19. I walked at 5 km/h for 6 hours. How far did I walk? ____

### Order of operations

Concept

Order
**1** ( )
**2** × and ÷
**3** + and −
(going from left to right)

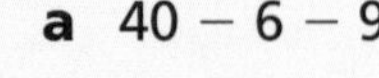

**a** $40 - 6 - 9$ ____
**b** $39 - 36 + 11$ ____
**c** $9 + 4 \times 6$ ____
**d** $16 - 4 \times 4$ ____
**e** $28 \div 7 \times 6$ ____
**f** $30 - (4 + 16)$ ____
**g** $16 - (2 \times 2)$ ____
**h** $50 - 6 \times 4$ ____
**i** $5 \times (39 - 36)$ ____
**j** $18 \div (56 - 50)$ ____
**k** $35 + 3 \times 5 - (49 - 9) \div (50 - 46)$ ____

 ISBN 978 0 6557 0886 5

## 31:3 ☐ out of 11

1. $4\overline{)84656}$
2. $7\overline{)30\cdot24}$
3. My painting time totalled 525 minutes over 3 days. On average, how long did I paint for each day? ______
4. The Olympic Games are held every 4 years . If they were held in 2000, will they be held in the future in 2065, 2066, 2067 or 2068? ______

5. Find the lowest common denominator, then find:
   a $\frac{9}{16} - \frac{1}{8} = \frac{\square}{\square} - \frac{\square}{\square} = \frac{\square}{\square}$
   b $\frac{1}{12} + \frac{1}{3} = \frac{\square}{\square} + \frac{\square}{\square} = \frac{\square}{\square}$
6. Use decimals to write these as litres.
   a 8000 mL = ______ L
   b 9465 mL = ______ L
   c 7 L 234 mL = ______ L
7. a 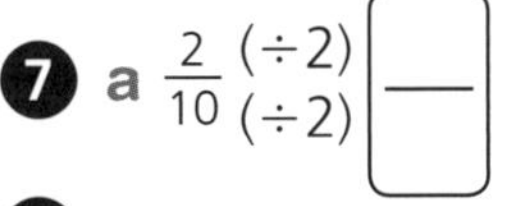 $\frac{2}{10}\,\frac{(\div 2)}{(\div 2)}$ ☐  b 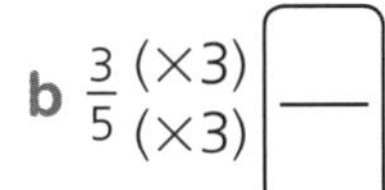 $\frac{3}{5}\,\frac{(\times 3)}{(\times 3)}$ ☐

8. Find the estimate by rounding each number to the nearest 10.
   a 19 × 56 = ______
   b 28 × 123 = ______
   c 498 × 31 = ______

9. Find the equal measures:
   a 8 L = ______ mL
   b 4839 mL = ______ L

10. 5 + (56 − 49) ÷ 2 ______
11. 13 children hold hands in a line. How many hands are being held? ______

## 31:4 ☐ out of 4

**Extension**

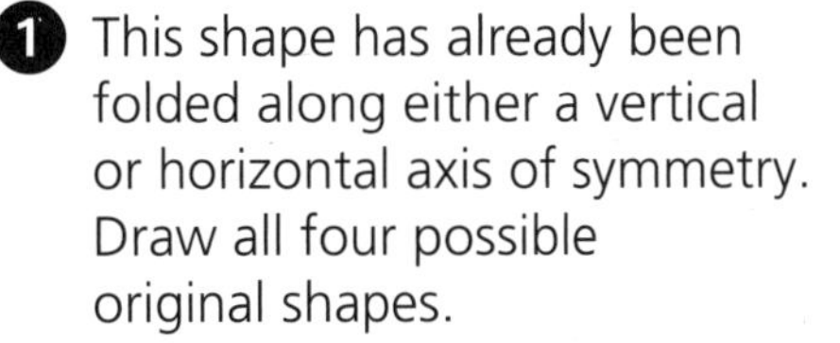

1. This shape has already been folded along either a vertical or horizontal axis of symmetry. Draw all four possible original shapes.

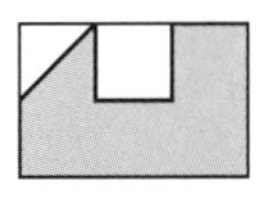

A
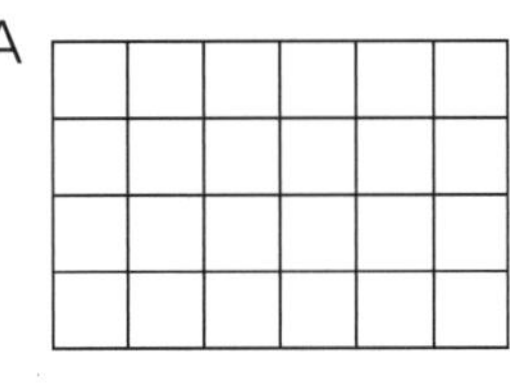

B
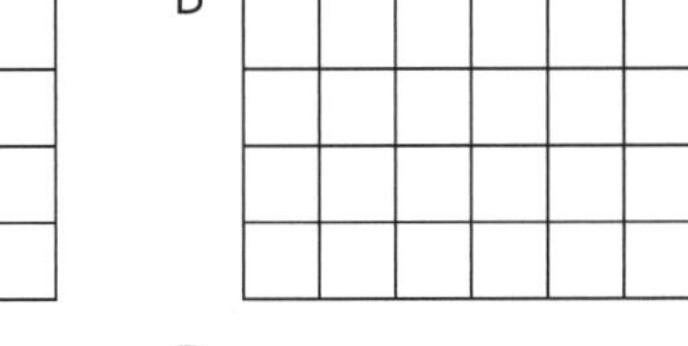

C
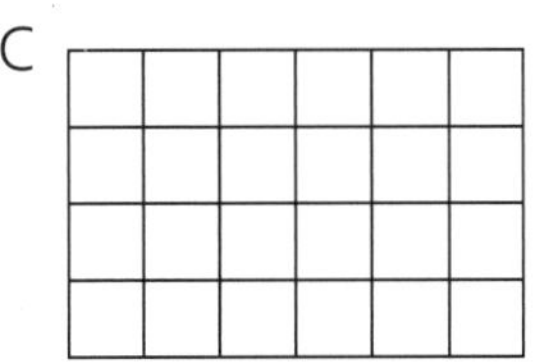

D
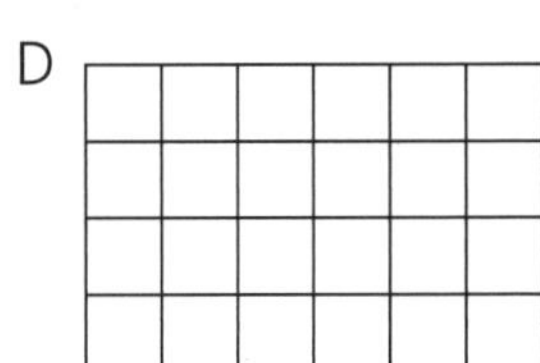

2. I read 45 minutes a night for 28 days. For how long did I read? ______
3. How many days are there until Christmas this year? ______
4. I take 7 steps forward, then 2 back, and then repeat this pattern. If I take one step every second, how many seconds will it take for me to go 14 steps forward from my starting position? ______

**Challenge**

*Record a time line for your life.*
*Fit in as many events and dates as you can.*

Year

Event  Birth

**Concept**

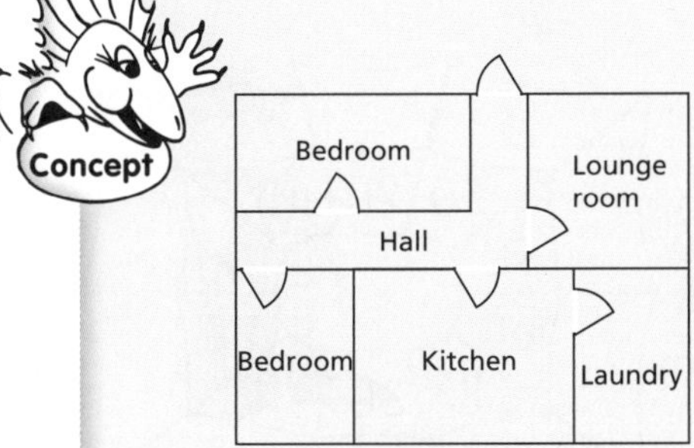


**Scale drawing**

Find the real length (the longest side) of:

a the laundry ______
b the lounge room ______
c the kitchen ______
d What is the real perimeter of the kitchen? ______

## 32:1 ☐ out of 16

1. $5 \times 6 + 7$ ____
2. $200 - 36$ ____
3. $8 \times 300$ ____
4. $9 \times 4 + 3$ ____
5. $\begin{array}{r} 9567{\cdot}3 \\ +\ \ 746{\cdot}4 \\ \hline \end{array}$
6. $\frac{9}{10} - \frac{4}{5}$ ____
7. $672 - 150$ ____
8. $\frac{1}{4}$ of 12 ____
9. $46 + 53 =$ ____ $+ 50$
10. $\begin{array}{r} 735{\cdot}07 \\ -\ \ 47{\cdot}93 \\ \hline \end{array}$

11.  
1 square tile
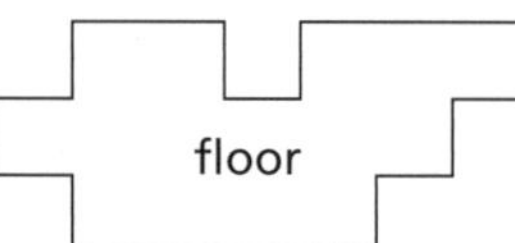

How many of the square tiles are needed to cover the floor? ____

12. $0{\cdot}7 = \frac{\square}{100} =$ ____ %

13. How many squares will there be if each side of the rectangle is made twice as long? ____
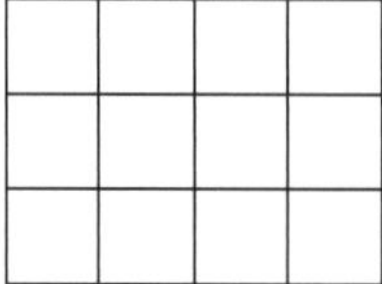

14. Find the lowest common denominator, then find:
 a $\frac{2}{10} + \frac{3}{4} = \frac{\square}{\square} + \frac{\square}{\square} = \frac{\square}{\square}$
 b $\frac{3}{4} + \frac{1}{3} = \frac{\square}{\square} + \frac{\square}{\square} = \frac{\square}{\square}$

15. Find the equal measures: (See page 85.)
 a 6·846 L = ____ mL
 b 7 L 278 mL = ____ mL
 c 8000 L = ____ kL

16. a $7 \times 70 =$ ____
 b $7 \times 7000 =$ ____
 c $50 \times 7000 =$ ____

## 32:2 ☐ out of 15

1. $5 \times 60$ L ____
2. $\frac{1}{3}$ of 21 ____
3. $49 \times 20$ ____
4. 50% of 18 ____
5. $\begin{array}{r} 35 \\ \times 76 \\ \hline \end{array}$ $(6 \times 35)$ ____ $(70 \times 35)$
6. 20% of $20 ____
7. $58 \times 300$ ____
8. $47 + 239 =$ ____ $+ 240$
9. $364 - 49 =$ ____ $- 50$
10. $\begin{array}{r} 93 \\ \times\ 67 \\ \hline \end{array}$ $(7 \times 93)$ ____ $(60 \times 93)$

11. If the sides of this rectangle are made 4 times as long, how many times as big will its area be? ____
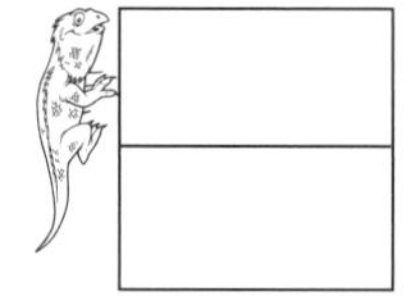

12. $100 \times 50 - 50 - 15 =$ ____

13. Find the lowest common denominator, then find:
 a $\frac{6}{7} - \frac{3}{10} = \frac{\square}{\square} - \frac{\square}{\square} = \frac{\square}{\square}$
 b $\frac{3}{10} + \frac{1}{3} = \frac{\square}{\square} + \frac{\square}{\square} = \frac{\square}{\square}$

14. Find the area of this shape.
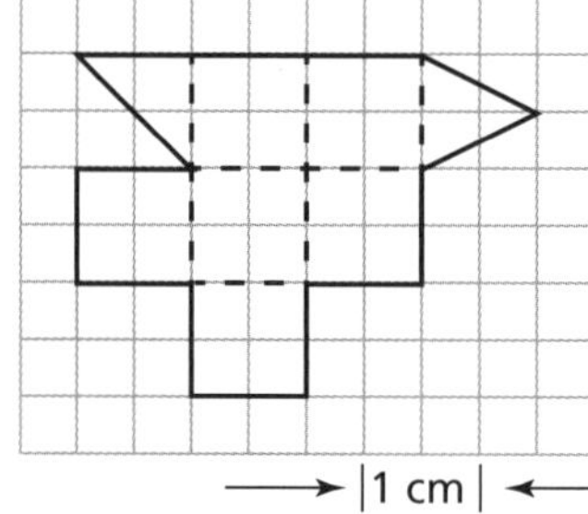

____

15. Use decimals to write these as litres.
 a 846 mL = ____ L
 b 58 mL = ____ L

× Tables

| ×6 | ×7 | ×8 |
|---|---|---|
| 10, 50, 80, 100, 40, 90, 60, 30, 70, 20 | 40, 50, 60, 100, 10, 90, 80, 30, 70, 20 | 50, 20, 40, 100, 80, 70, 30, 10, 60, 90 |



 ISBN 978 0 6557 0886 5

## 32:3 out of 10

1. $6\overline{)38094}$

2. $3\overline{)23{\cdot}61}$

3. 
```
  23
 ×45
 ___ (5 × 23)
     (40 × 23)
 ___
```

4. 
```
  86
 ×67
 ___ (7 × 86)
     (60 × 86)
 ___
```

5. Find the lowest common denominator, then find:
   a $\frac{8}{12} - \frac{1}{4} = \frac{\square}{\square} - \frac{\square}{\square} = \frac{\square}{\square}$
   b $\frac{3}{4} + \frac{1}{10} = \frac{\square}{\square} + \frac{\square}{\square} = \frac{\square}{\square}$

6. Use decimals to write these as litres.
   a 298 mL = ______ L
   b 46 mL = ______ L
   c 7 mL = ______ L

7. Daniel was paid $20 an hour for 78 hours of work. How much was he paid altogether? ______

8. Find the estimate by rounding each number to the nearest 10.
   a 21 × 99 = ______
   b 48 × 121 = ______
   c 504 × 21 = ______

9. Find the equal measures:
   a 15 000 L = ______ kL
   b 4 kL = ______ L
   c 76·395 kL = ______ L
   d 3 ML = ______ kL
   e 57 684 kL = ______ ML

10. (76 − 35) + 8 × 4 = ______

## 32:4 out of 5

**Extension**

1. a How many small squares are in this rectangle? ______
   b How many are in an area 4 times as long and 3 times as wide? ______

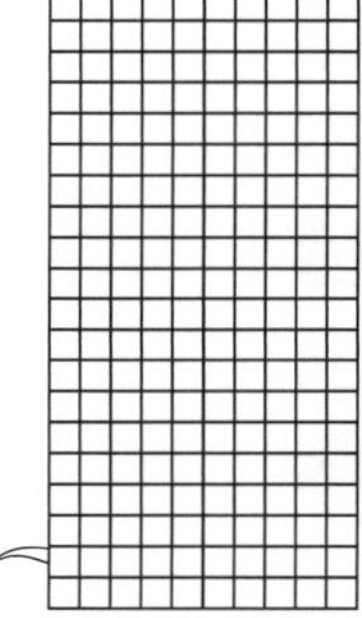

2. I spent $\frac{1}{4}$ of my money on a t-shirt, $\frac{1}{3}$ of my money on pants and $\frac{1}{6}$ of my money on lunch. I had $18 left. How much money did I start with? ______

3. Rachel swims 26 laps and Marika swims 33 laps of the 50 m pool every Thursday. How much further does Marika swim on:
   a one visit? ______
   b 8 visits? ______

4. Jody started with a number, doubled it, divided it by 3 then subtracted 52. She finished with 8. What did she start with? ______

5. To the total of 14, 18 and 38, add the product of 45 and 167. ______

**Challenge**

*Research containers that hold megalitres and list each name and approximate capacity below.*
*e.g. An Olympic swimming pool holds* ______

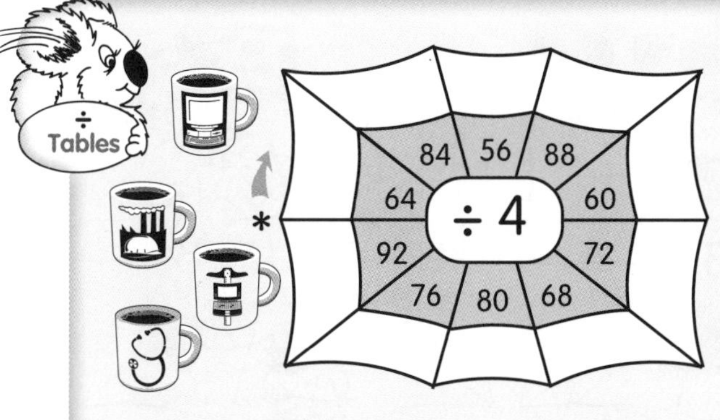


**To divide numbers by 4, halve the number, then halve it again.**

a 72 students were put into 4 equal teams. How many were put in each team? ______

b How many rows of 4 can be made with 96 mugs? ______

c Naomi, Alana, Luke and Heather shared 128 grapes. How many did each receive? ______

 • *AUSTRALIAN SIGNPOST MATHS 6 MENTALS* • ISBN 978 0 6557 0886 5

## 33:1

out of 16

1. $8 \times 9 + 14$ ____
2. $600 - 39$ ____
3. $5 \times 600$ ____
4. $7 \times 6 \div 7$ ____
5. $67 \times 8$ ____
6. $\frac{2}{6} + \frac{4}{10}$ ____
7. 20% of \$40 ____
8. $6{\cdot}434$ m $\times$ 1000 ____
9. $3 \times 800$ L ____
10. $42 \times 9$ ____

11. a 5·926 L = ________ mL

    b 2 L 978 mL = ________ mL

    c 6000 L = ________ kL

12. How much further must this car travel before the odometer shows 1000 km?

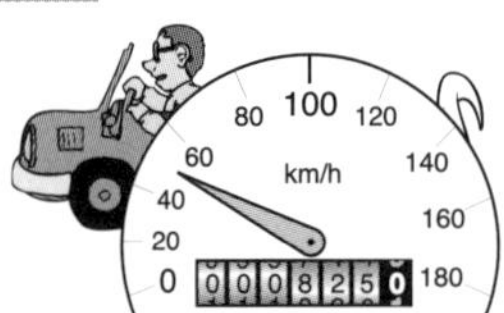


________

13. Underline the units that have been correctly written.

KG G gm g. Kg
g k.g. kg G. kG

14. a The abbreviation for kilolitres is ____.

    b The abbreviation for megalitres is ____.

    c The abbreviation for tonnes is ____.

15. Find the equal measures: (See page 85.)

    a 6 t = ________ kg

    b 3000 kg = ________ t

16. a Is $\frac{2}{10}$ the same as $\frac{1}{5}$? ____

    b Is $\frac{5}{10}$ the same as $\frac{1}{2}$? ____

    c $\frac{2}{10} + \frac{5}{10}$ ____

    d $\frac{1}{5} + \frac{1}{2}$ ____

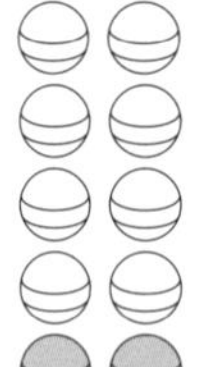

## 33:2

out of 15

1. $473 + 387$ ____
2. $894 - 354$ ____
3. $69 - 56 \div 8$ ____
4. $\frac{1}{3}$ of \$27 ____
5. $59 \times 46$ $(6 \times 59)$ $(40 \times 59)$ ____
6. $\frac{7}{10} + \frac{1}{3}$ ____
7. \$20 – \$6.73 ____
8. $376 - 89 =$ ____ $- 90$
9. $437 + 38 =$ ____ $+ 40$
10. $71 \times 53$ $(3 \times 71)$ $(50 \times 71)$ ____

11. $(7 \times 10^4) + (8 \times 10^3) + (2 \times 10^2) + (9 \times 10^1) + 1$

    = ________

12. Lachlan's Nanna gave \$5 for every goal scored in the soccer season.

    a If he scored 36 goals, how much money did he receive? ________

    b If his sister Felicity scored 58, how much did she receive? ________

13. Use decimals to write these as litres.

    a 298 mL = ________ L

    b 84 mL = ________ L

    c 19 mL = ________ L

14. a 7·4 t = ________ kg

    b 90 000 kg = ________ t

15. 

Each dog has a mass of about 15 kg. Calculate the approximate mass of:

a 5 dogs ________ b 14 dogs ________

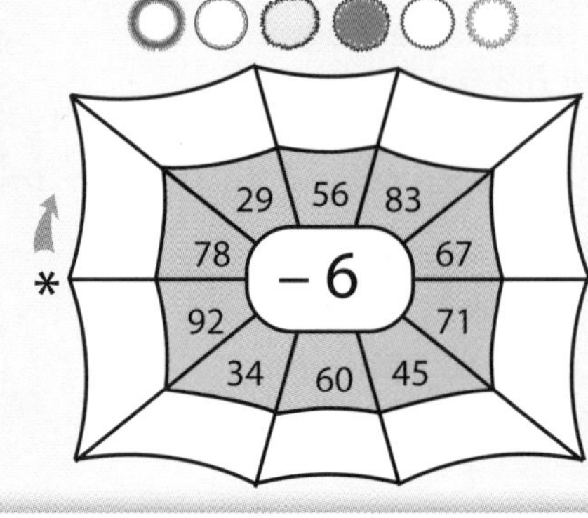

 ISBN 978 0 6557 0886 5

## 33:3 ☐ out of 8

1.
```
   8 3
 × 7 6
 _____ (6 × 83)
 _____ (70 × 83)
 _____
```

2.
```
   9 2 6
 ×   7 8
 _______ (8 × 926)
 _______ (70 × 926)
 _______
```

3. a A whale weighs 140 000 kg. How many tonnes is this? ______
   b A car had a mass of 1·8 t. How many kg is this? ______

4. a 2·97 t ______ kg  b 30 000 kg ______ t

5. a Write this as 24-hour time. ______
   b What will be the 24-hour time in 18 minutes? ______

6. Round each number to the nearest whole, then estimate the answer.
   a 10·2 × 5·9 = ______
   b 12·5 + 29·8 = ______
   c 52·6 − 30·5 = ______
   d 42 × 3·4 = ______

7. Use decimals to write these as litres.
   a 289 mL = ______ L
   b 37 mL = ______ L
   c 6 mL = ______ L

See page 85 for help.

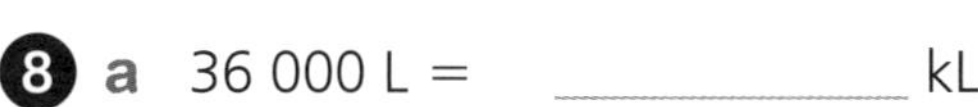

8. a 36 000 L = ______ kL
   b 9 kL = ______ L
   c 27·746 kL = ______ L
   d 7 ML = ______ kL
   e 27 453 kL = ______ ML

## 33:4 ☐ out of 4

**Extension**

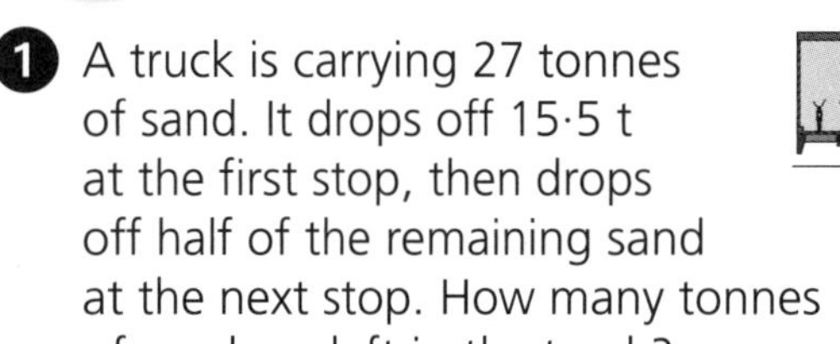

1. A truck is carrying 27 tonnes of sand. It drops off 15·5 t at the first stop, then drops off half of the remaining sand at the next stop. How many tonnes of sand are left in the truck? ______

2. Draw the missing part of this pattern.

3. Tanya and Karyn took 60 cakes each to the cake stall. Tanya sold $\frac{9}{12}$ of her cakes and Karyn sold $\frac{11}{12}$ of her cakes.
   a How many cakes did they sell altogether? ______
   b How many did they have left? ______

4. If you double the length, width and height of this shape, how many times as big will its volume be?

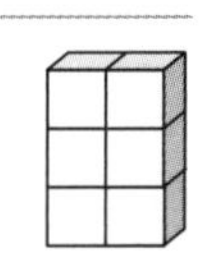

______

**Challenge**

*Research objects that are measured in tonnes. List each below with an approximate mass if appropriate, e.g. Semitrailer ______.*

______

______

______

______

______

**Concept**

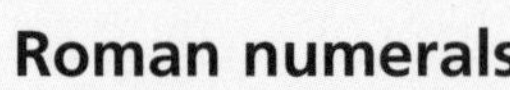

### Roman numerals

| 1 | I | one finger |
|---|---|---|
| 5 | V | one hand |
| 10 | X | two Vs |
| 50 | L | half of a C |
| 100 | C | centum = 100 |
| 500 | D | half of an ⓘ |
| 1000 | M | mille = 1000 |

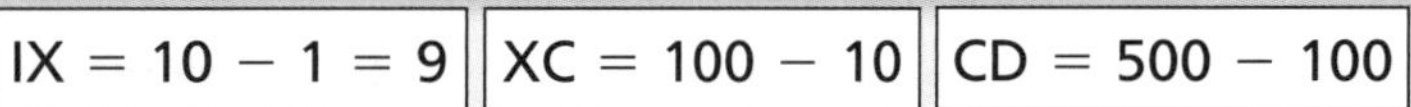

IX = 10 − 1 = 9 | XC = 100 − 10 | CD = 500 − 100

1. Write our numeral for:
   a XCVI ______  b CLX ______
   c DCX ______  d CCIV ______
   e LXVI ______  f XXXVI ______

2. Write the Roman numeral for:
   a 75 ______  b 380 ______
   c 378 ______  d 2567 ______
   e 635 ______  f 3999 ______

## 34:1   out of 17

1. ____ + 8 = 10
2. 18 + ____ = 30
3. $\frac{11}{12} + \frac{5}{6}$ ____
4. $7 \times 6 - 12$ ____
5. $87 \times 6$ ____
6. 40% of $80 ____
7. 8·475 m × 1000 ____
8. 634 – 213 ____
9. $6 – $2.50 ____
10. $21 \times 32$ ____

11. Find the equal measures: (See page 85.)

| Tonnes | Kilograms |
|---|---|
| 7t | |
| | 70000 kg |
| 7·546 t | |

| Litres | Millilitres |
|---|---|
| 9L | |
| | 2000 mL |
| 8·3 L | |

12. < or >? **a** $\frac{8}{10}$ ☐ $\frac{3}{5}$   **b** $\frac{2}{6}$ ☐ $\frac{2}{3}$

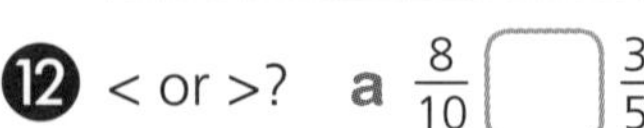

13. 192 – ☐ = 65   ☐ = ____
14. mg is short for ____________.
15. How many pairs of parallel edges has this triangular prism?

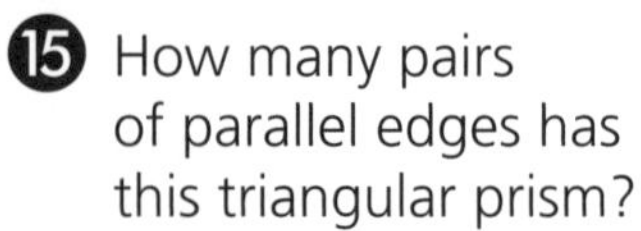

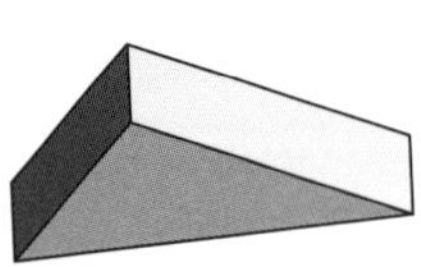

____________

16. What is the best unit to measure the mass of a:
    - **a** full suitcase? ____________
    - **b** grain of sand? ____________
    - **c** shoe? ____________
17. Name this solid. ____________

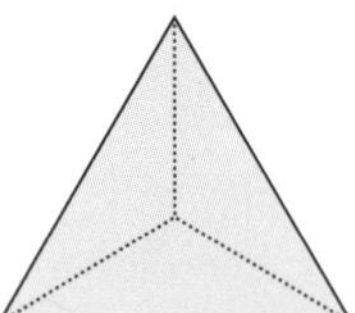

## 34:2   out of 17

1. ____ + 47 = 100
2. $\frac{4}{7}$ of 63 ____
3. 56 ÷ ____ = 8
4. $\frac{1}{2} + \frac{5}{12}$ ____
5. $42 \times 89$ ____
6. $27 – $5·60 ____
7. Triple 325. ____
8. 6 × 5 = 57 – ____
9. ____ – 142 = 4 × 8
10. $836 \times 57$ ____

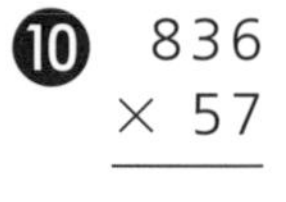

11. $2{\cdot}5 \times 4\ 6$ ____
12. $3{\cdot}9 \times 5\ 7$ ____
13. $3{\cdot}4 \times 7\ 4$ ____
14. **a** $11\frac{1}{2} + 9$ ____________
    **b** $7\frac{1}{2} + 3\frac{1}{2}$ ____________

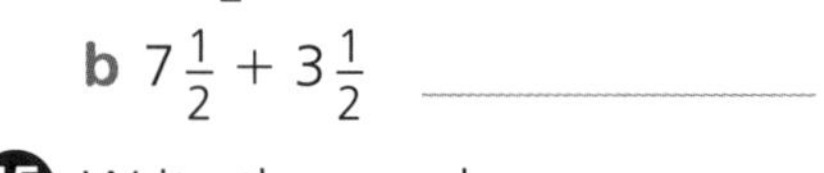

15. Write the equal masses.

| Tonnes & kilograms | Tonnes |
|---|---|
| 7 t 300 kg | |
| | 9·546 t |
| 5 t 476 kg | |

16. 17 – ☐ = 47 – 39   ☐ = ____
17. **a** 4768 g = ____________ kg
    **b** 4 g = ____________ mg
    **c** 67 t = ____________ kg
    **d** 46·8 kg = ____________ g
    **e** 10 mg = ____________ g

Concept

**Scale: 1 cm represents 50 cm**

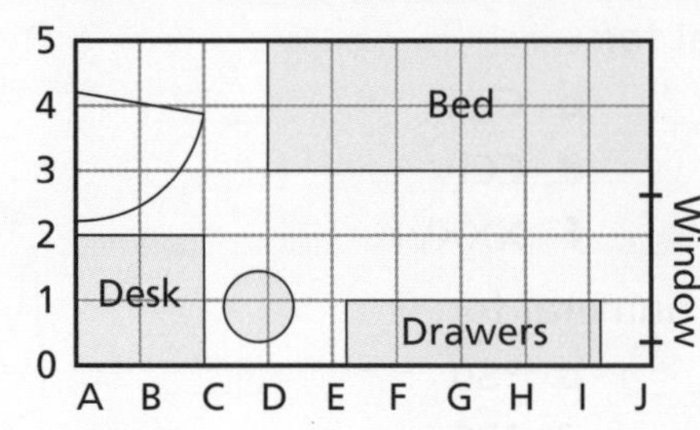


**Scale drawing**

**a** The scale of this drawing ____ : ____

What is the actual length of:

**b** the bed? ____

**c** the drawers? ____

What is at the coordinate position:

**d** G4? ____________   **e** B1? ____________

**f** Which coordinates could be used for the window? ____



 *AUSTRALIAN SIGNPOST MATHS 6 MENTALS* • ISBN 978 0 6557 0886 5

## 34:3

out of 15

1. $35 \times 68$
2. $73 \times 18$
3. $384 \times 46$
4. $5{\cdot}9 \times 13$
5. $8{\cdot}4 \times 25$
6. $9{\cdot}8 \times 55$
7. The mass of 3 trucks is 9·1 t. What is the mass of:
   a 6 trucks?
   b 30 trucks?

8. Complete: $\frac{1}{2} = \frac{2}{4} = \frac{\square}{8} = \frac{5}{\square}$
9. a 9·1 t = ______ kg
   b 40 000 kg = ______ t

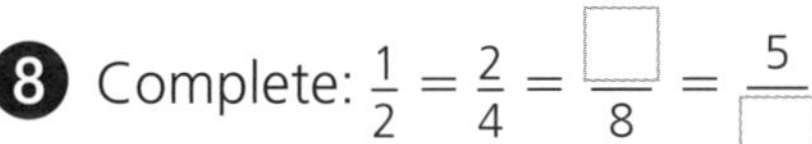

10. 1000 − 37 − 37 − 37 − 37
11. What is the total mass, in kilograms, of a bag of 9 potatoes, each with an average mass of 326 g?
12. 9 × 2 + □ = 25 □ = ______
13. a 4 g = ______ mg
    b 10 mg = ______ g
14. 600 children were each given $20. How much money was given altogether?
15. Complete the table if: △ = □ × 46

| □ | 20 | 13 | 26 |
|---|---|---|---|
| △ | | | |

## 34:4

Extension

out of 6

1. A boat weighs the same as 46 people. How many people would weigh the same as 12 boats?
2. Write all possible pairs of counting numbers that make this number sentence true.
   2707 − 2635 = △ × □
3. My trolley contained three 1700 g bags of potatoes and a pumpkin weighing 3·56 kg.
   a What is the total mass of my groceries?
   b How much short of 10 kg is the mass of my groceries?
4. If an ant's mass is 3·75 mg, how much could it carry if it can carry 35 times its weight?

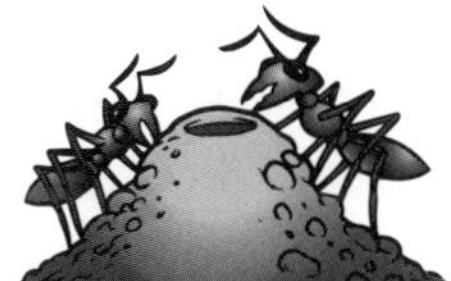

5. 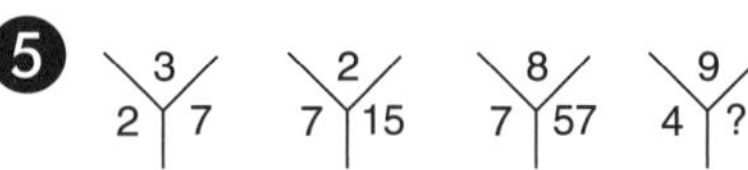


   What is the missing number?
6. 10 000 − 256 − 256 − 256 − 256

**Challenge**

*Complete:*

a △ + 7 = 16 so △ = ______
b ▱ − 62 = 23 so ▱ = ______
c 2 × △ = 98 so △ = ______
d ▱ ÷ 9 = 7 so ▱ = ______
e △ + 84 = 106 − 9 so △ = ______
f 8 × ▱ = 45 + 19 so ▱ = ______

**Coordinates**

(1R, 3U) is the point 1 right and 3 up from zero. Join the coordinates in order, to complete the picture.

(1R, 3U), (3R, 5U), (8R, 5U), (10R, 4U), (12R, 6U), (12R, 1U), (10R, 3U), (8R, 2U), (3R, 2U), (1R, 3U)

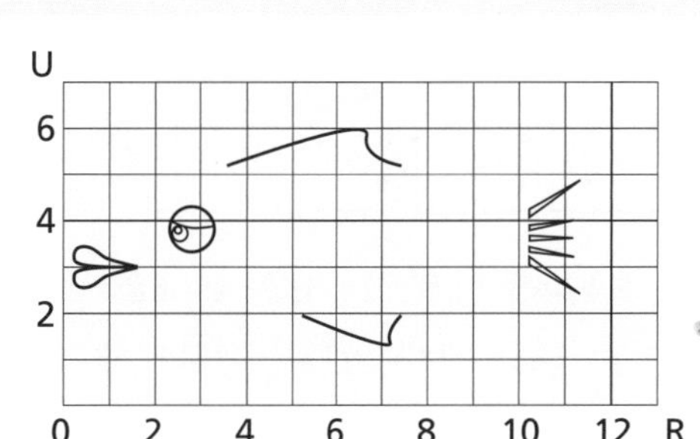


  ISBN 978 0 6557 0886 5

## 35:1 ☐ out of 18

1. 100 − 24 ____
2. 10% of 60 ____
3. 7 × ____ = 56
4. $\frac{3}{4} - \frac{1}{2}$ ____
5. $\begin{array}{r} 87 \\ \times \ \ 6 \\ \hline \end{array}$
6. 78 + 39 = ____ + 40
7. 35 − ____ = 5 × 6
8. 5·498 L × 1000 ____
9. \$5 × 80 ____
10. $\begin{array}{r} 28 \\ \times 11 \\ \hline \end{array}$ ____
11. 128 ÷ ☐ = 32 ☐ = ____
12. a 7 t = ____ kg
    b 8 000 kg = ____ t
    c 0·276 kg = ____ g
    d 7 kg 925 g = ____ g
    e 200 ha = ____ $km^2$

13. Which of +, − , × and ÷ will make 3 ☐ 3 △ 3 = 4 true? ☐ = ____, △ = ____
14. Complete this pattern:

| Decagons | 1 | 2 | 3 | 4 | 5 |
|---|---|---|---|---|---|
| Sides | 10 | 20 | | | |

15. a $\frac{8}{10}\frac{(\div 2)}{(\div 2)} = \frac{☐}{☐}$ b $\frac{1}{4}\frac{(\times 3)}{(\times 3)} = \frac{☐}{☐}$
16. a A prime number has ____ factors, 1 and ____.
    b A composite numbers has more than ____ factors.
17. How many halves are in 8 apples? ____
18. What is the abbreviation for:
    a hectares? ____
    b square kilometres? ____

## 35:2 ☐ out of 18

1. 453 + 646 ____
2. 684 − 298 ____
3. $\frac{1}{6} + \frac{7}{10}$ ____
4. $10 - 8\frac{2}{5}$ ____
5. $\begin{array}{r} 77 \\ \times 37 \\ \hline \end{array}$ ____
6. $\frac{2}{3}$ of 27 ____
7. 25% of 50 ____
8. 7 × (67 − 62) ____
9. $\frac{4}{6} - \frac{1}{2}$ ____
10. $\begin{array}{r} 453 \\ \times \ 34 \\ \hline \end{array}$ ____
11. $\begin{array}{r} 2·8 \\ \times 1\ 5 \\ \hline \end{array}$ ____
12. $\begin{array}{r} 7·6 \\ \times 6\ 7 \\ \hline \end{array}$ ____
13. $\begin{array}{r} 7·5 \\ \times 8\ 8 \\ \hline \end{array}$ ____
14. a 7·76 t = ____ kg
    b 20 000 kg = ____ t
    c 0·456 kg = ____ g
    d 2 kg 487 g = ____ g
    e 9 g = ____ mg
    f 10 mg = ____ g
15. A bag of carrots weighs 0·427 kg. If there are 7 carrots in the bag, what is the average mass of each carrot?

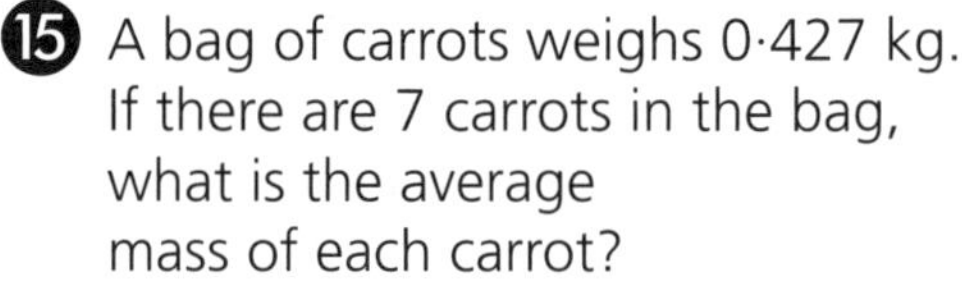

16. (3 + △) × 7 = 35 △ = ____
17. Circle the prime numbers.
    17 21 29 46 83 67 49
18. a 200 ha = ____ $km^2$
    b 60 000 $m^2$ = ____ ha
    c 500 ha = ____ $km^2$
    d 9 $km^2$ = ____ ha

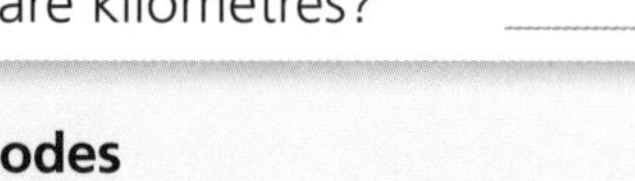

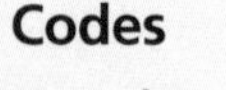


### Codes

Use the code to write "true" or "false" for each code statement.

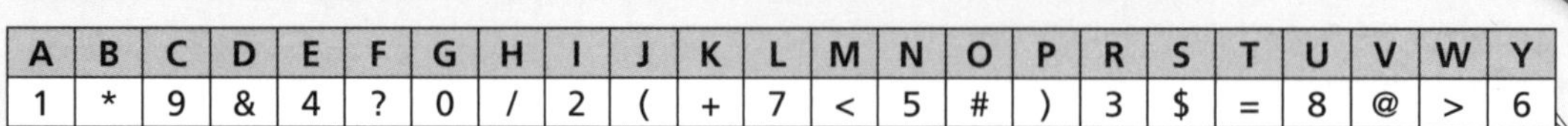

| A | B | C | D | E | F | G | H | I | J | K | L | M | N | O | P | R | S | T | U | V | W | Y |
|---|---|---|---|---|---|---|---|---|---|---|---|---|---|---|---|---|---|---|---|---|---|---|
| 1 | * | 9 | & | 4 | ? | 0 | / | 2 | ( | + | 7 | < | 5 | # | ) | 3 | $ | = | 8 | @ | > | 6 |

a \$4@45 2\$ 1 )32<4 58<*43 ____
b 1 )45=10#5 /1\$ ?2@4 \$2&4\$ ____
c 420/= 2\$ #54 ?19=#3 #? =>45=6 ?#83 ____


 • *AUSTRALIAN SIGNPOST MATHS 6 MENTALS* • ISBN 978 0 6557 0886 5

## 35:3 ☐ out of 9

1. $\begin{array}{r} 73 \\ \times\ 29 \\ \hline \end{array}$

2. $\begin{array}{r} 5{\cdot}7 \\ \times\ 36 \\ \hline \end{array}$

3. $\begin{array}{r} 381 \\ \times\ 23 \\ \hline \end{array}$

4. $\frac{3}{4} + \frac{1}{8} = \frac{\square}{\square} + \frac{\square}{\square} = \frac{\square}{\square}$

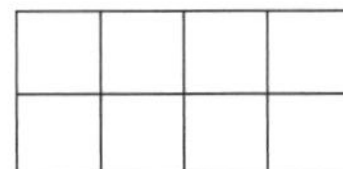

5. **a** 4·34 t = ______ kg
   **b** 50 000 kg = ______ t
   **c** 0·907 kg = ______ g
   **d** 5 kg 264 g = ______ g
   **e** 4 g = ______ mg

6. **a** ☐ + 56 = 93 ☐ = ____
   **b** ☐ − 13 = 47 ☐ = ____
   **c** ☐ × 8 = 992 ☐ = ____

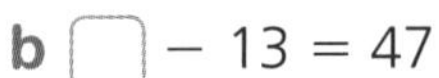

7. The mass of a horse is 0·7 tonnes. Find the mass of:
   **a** 18 horses ______
   **b** 38 horses ______

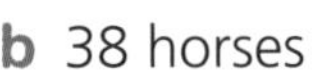

8. The mass of 10 men is 0·8 t. What is the mass of:
   **a** 30 men? **b** 50 men?

9. A square with 100 m sides has an area of 1 hectare. (100 m × 100 m = 1 ha). Find the area (in ha) of each of these.

   **a** (rectangle) 100 m, 200 m

   **b** (square) 200 m, 200 m

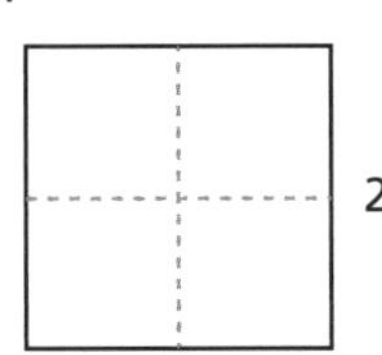

## 35:4 ☐ out of 4

**Extension**

1. Hume dam has a capacity of 3 005 157 ML. Blowering dam has a capacity of 1 628 000 ML.
   **a** If there is currently 2 857 214 ML in Hume dam, how much more water could it hold before it reaches its capacity? ______
   **b** If there is currently 876 978 ML in Blowering dam how much more water could it hold? ______

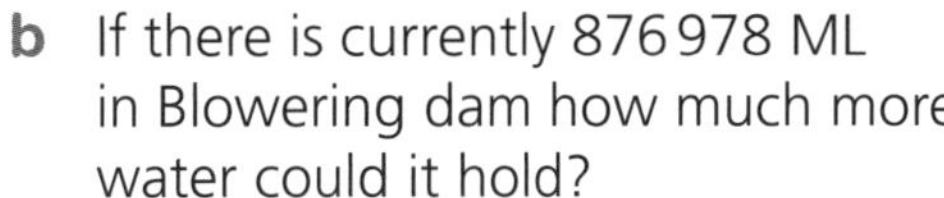

2. What number is missing? ______

   | 137 | 217 | 272 | ? |
   |---|---|---|---|
   | 89 \| 45 | 145 \| 69 | 172 \| 97 | 198 \| 134 |

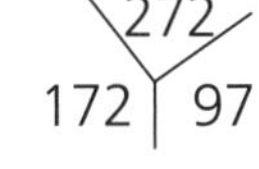
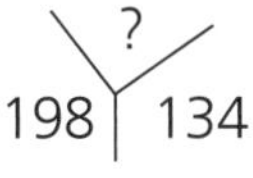

3. **a** $\begin{array}{r} 6\cdot\square\square 7 \\ +\ \square\cdot 0\ 4\ \square \\ \hline 14\cdot 9\ 7\ 6 \end{array}$ **b** $\begin{array}{r} 8\cdot\square 5\ \square \\ +\ \square\cdot 8\ \square\ 8 \\ \hline 16\cdot 6\ 3\ 0 \end{array}$

4. A male fly weighs about 11·5 mg and a female fly weighs about 17·5 mg. What is the difference between the mass of 6 male flies and 6 female flies? ______

**Challenge**

*Research objects that are measured in milligrams. List each below with an approximate mass, e.g. A grain of rice weighs* ______

**Example of factors**

Draw the rectangles that have an area of 8 units².

Draw rectangles of area 15 units².

List the factors. ______

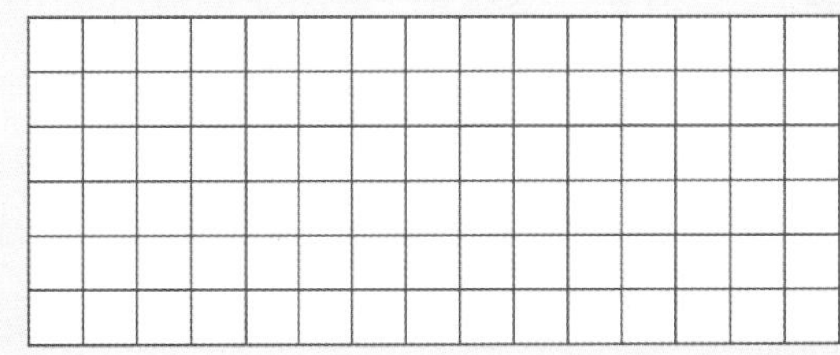

## 36:1 out of 15

1. 409 + 356 ____
2. 600 − 57 ____
3. \$9 × 400 ____
4. 20% of 40 ____
5. $\begin{array}{r} 375{\cdot}89 \\ +\ 857{\cdot}46 \\ \hline \end{array}$
6. 7 × ____ = 47 − 40
7. 75 − 35 ____ − 40
8. 8 + 4 ÷ 2 ____
9. $\frac{6}{10} - \frac{1}{2}$ ____
10. $\begin{array}{r} \$9786{\cdot}37 \\ -\ \$978{\cdot}36 \\ \hline \end{array}$

11. Circle the composite numbers.

5 9 12 17 11 6 15

12. a $\frac{6}{10}\ \frac{(\div 2)}{(\div 2)}$ $\frac{\square}{\square}$   b $\frac{1}{8}\ \frac{(\times 3)}{(\times 3)}$ $\frac{\square}{\square}$

13. List the factors for each number.
Put a tick next to the prime numbers.

| 1 | | | |
|---|---|---|---|
| 2 | | | |
| 3 | | | |
| 4 | | | |

| 5 | | | | |
|---|---|---|---|---|
| 6 | | | | |
| 7 | | | | |
| 8 | | | | |

14. a $\frac{3}{4}$ of 28 ____
b $\frac{1}{7}$ of 28 ____
c $\frac{3}{7}$ of 28 ____
d $\frac{5}{7}$ of 28 ____

15.

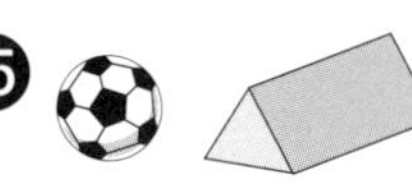

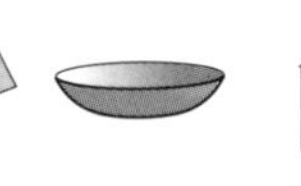

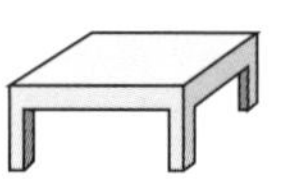

A B C D E

Which have a top view that is:

a a circle? ____
b a rectangle? ____

## 36:2

out of 17

1. 60 ÷ 2 + 54 ____
2. $8^2 + 2^2$ ____
3. $\frac{3}{5}$ of 40 ____
4. $\frac{8}{10} - \frac{2}{3}$ ____
5. $\begin{array}{r} 50{\cdot}9006 \\ -\ \ 7{\cdot}5673 \\ \hline \end{array}$
6. 6 × 7 = ____ − 58
7. 90 + 3 × 15 − 8 ____
8. $6 - \frac{5}{6}$ ____
9. 5·624 m × 1000 ____
10. $\begin{array}{r} 900{\cdot}004 \\ -\ \ 46{\cdot}896 \\ \hline \end{array}$

11. a 700 ha = ____ $km^2$
b 20 000 $m^2$ = ____ ha
c 900 ha ____ $km^2$
d 5 $km^2$ = ____ ha

12. 7 + Δ × 2 = 10 + 5   Δ = ____

13. $\frac{7}{10} - \frac{1}{5} = \frac{\square}{\square} - \frac{\square}{\square} = \frac{\square}{\square}$

14. Circle the numbers that are divisible by 5.
(The last digit must be 5 or 0.)

852 790 555 231 685 917 080

15. Circle the prime numbers.

19 45 31 69 47 51 62

16. List all the factors of 42.

____

17.

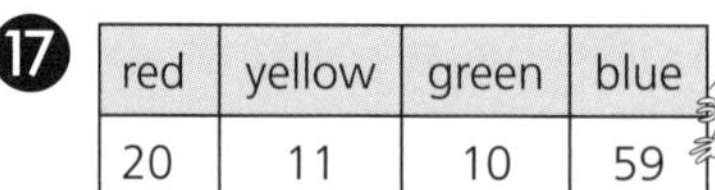

| red | yellow | green | blue |
|---|---|---|---|
| 20 | 11 | 10 | 59 |

This table shows the results from a survey of 100 children in my school. Students were asked what their preferred colour was. If you asked 10 students their preferred colour, predict how many would say:

a red? ____   b green? ____

Which vowel is used the most?

a Which vowel do you think is used the most? ____

b Use this tally to record the vowels in this activity.

| | a | e | i | o | u |
|---|---|---|---|---|---|
| Tally | | | | | |

c Which vowel was used the most? ____

## 36:3 out of 11

1. 
```
  4572·94
   835·08
     2·45
+  845·37
```

2. 
```
  $9678.25
− $3656.93
```

3. a I took one hour to do my homework. Alana took half as long. Heather took one third as long as Alana. How long did Heather take? ______
   b In part a, how much time was spent on homework altogether? ______

4. 100 − 7 − 7 − 7 − 7 − 7 ______

5. ☐ × 4 = 200 ☐ = ______

6. The bath's water level was 21·3 cm before Archimedes got in and 25·7 cm after. What was the difference in water level? ______

7. List all the prime numbers less than 30.

______

8. a 800 ha = ______ km²
   b 40 000 m² = ______ ha
   c 900 ha ______ km²
   d 12 km² = ______ ha

9. List the factors for each number. Put a tick next to the prime numbers.

| 9 | | | | | | |
|---|---|---|---|---|---|---|
| 10 | | | | | | |
| 11 | | | | | | |
| 12 | | | | | | |

| 13 | | | | | |
|---|---|---|---|---|---|
| 14 | | | | | |
| 15 | | | | | |
| 16 | | | | | |

10. Circle the numbers that are divisible by 10.

2307 800 010 410 043 8310 7850

11. List all the factors of 100.

______

## 36:4 Extension out of 7

1. Write the missing numbers in the algorithms.

a
```
    ☐5☐
×    ☐8
  60 7 2
 227 7 0
 ☐
```

b
```
     ☐
×   54
  1696
 21200
 ☐
```

c
```
     ☐
×   39
  2664
  8880
 ☐
```

2. 30 000 − 89 − 89 − 89 − 89 ______

3. Draw the missing part of this pattern.

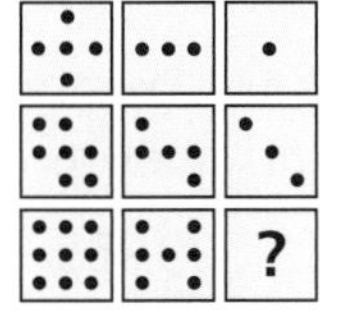

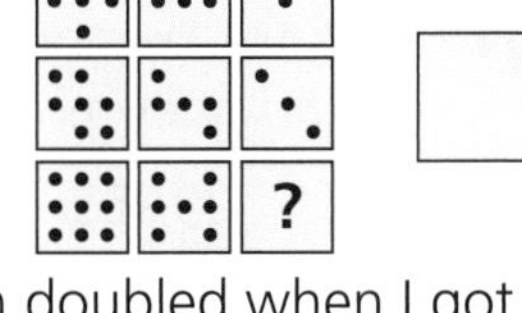

4. The water level of the bath doubled when I got in, then I added water and the water level rose 8·34 cm to 21·8 cm. What was the water level before I got in? ______

5. The shaded part has a value of 48.
   a What is the value of the whole? ______
   b What is the value of $7\frac{1}{3}$ whole triangles like this? ______

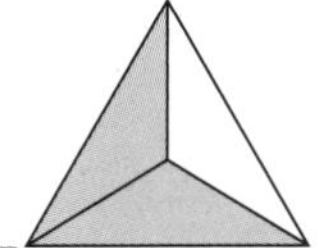

6. 1000 − 19 − 19 − 19 − 19 − 19 = ______

7. $1^2 + 2^2 + 3^2 + 4^2 + 5^2 + 6^2 + 7^2$ ______

**Challenge**

*Survey 20 people. Ask them how many siblings (brothers and sisters) they have. Tally the results in the table.*

| Siblings | | Total |
|---|---|---|
| 0 | | |
| 1 | | |
| 2 | | |
| 3 | | |
| 4 | | |
| other | | |

### Testing divisibility rules using a calculator

a Choose any 10 large numbers ending in 0 or 5. Is each one divisible by 5? ______

b Choose any 10 large numbers whose last two digits are divisible by 4 (like 810 912). Is each one divisible by 4? ______

c Choose any 10 large numbers whose sum of digits is divisible by 3 (like 411 441). Is each one divisible by 3? ______

One number is divisible by another if the answer is a whole number when you divide.

## 37:1 — out of 15

1. 0·2 + 0·6 ____
2. 20% of 80 ____
3. 5 + 14 ÷ 2 ____
4. $\frac{3}{8} + \frac{1}{2}$ ____
5. 
```
  364700
− 48678
```
6. \$7 − \$2.80 ____
7. 0 4 − 0 3 ____
8. 8 345 × 1000 ____
9. 57 + ____ = 100
10. 
```
  490080
− 37567
```
11. Circle the numbers that are divisible by 2. (The number must be even.)

    6578  29053  6475  2080  35245

12. **a** What ingredient is used most in making a quiche? ____

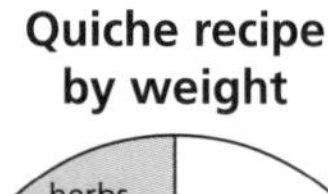


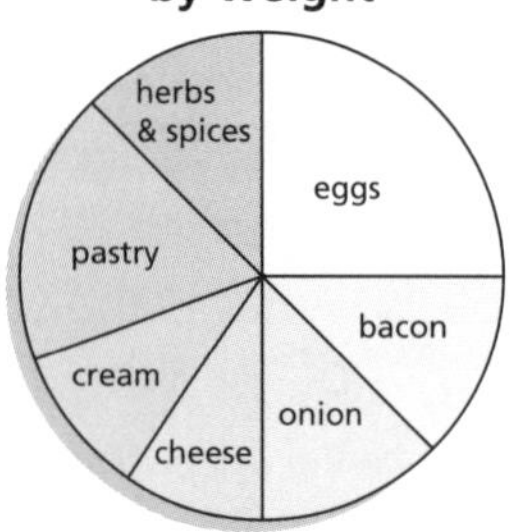


**b** What fraction of the final quiche is made from eggs? ____

**c** If the quiche weighs 400 g, how many 50 g eggs were used in the recipe? ____

Which is used more:

**d** bacon or cheese? ____

**e** pastry or onion? ____

13. If the rule is F + 5 what will the answer be if F is:

    **a** 7? ____ **b** 16? ____

14. My car can travel 12 km on one litre of petrol. How far can I go with four litres? ____

15. The value of the 3 in 78·943. ____

## 37:2 — out of 14

1. 2·8 + 0·5 ____
2. 1·5 ÷ 0·5 ____
3. 45 − 4 × 7 ____
4. $8\frac{7}{8} - \frac{1}{2}$ ____
5. 
```
  $ 500·00
+ $ 692·49
```
6. $\frac{3}{8}$ of 64 ____
7. 893 − 536 ____
8. 100 − ____ = 6 × 7
9. $\frac{7}{10} + \frac{8}{5}$ ____
10. 
```
  $7500·00
− $  936·93
```
11. List all the factors of 80. ____

12. 

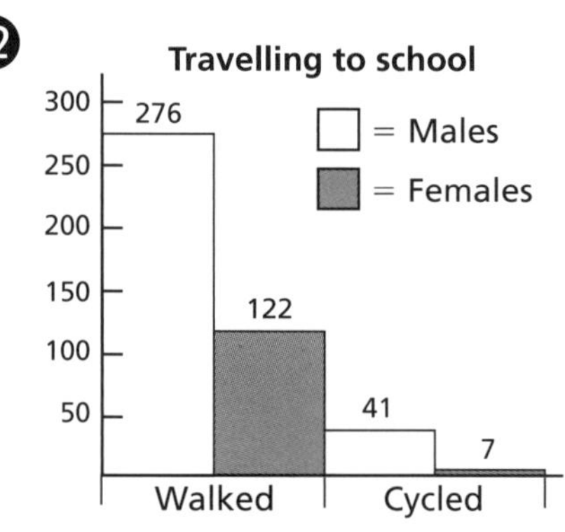

**a** How many people who walked were females? ____

**b** Is it more likely that the next person to arrive at school by walking will be male or female? ____

**c** How many people walked altogether? ____

13. Circle the numbers that are divisible by 4. (The number made by the last 2 digits must be divisible by 4.)

    36  100  812  561  4000

14. $(5 \times 10^4) + (6 \times 10^3) + (4 \times 10^2) =$ ____

## Concept

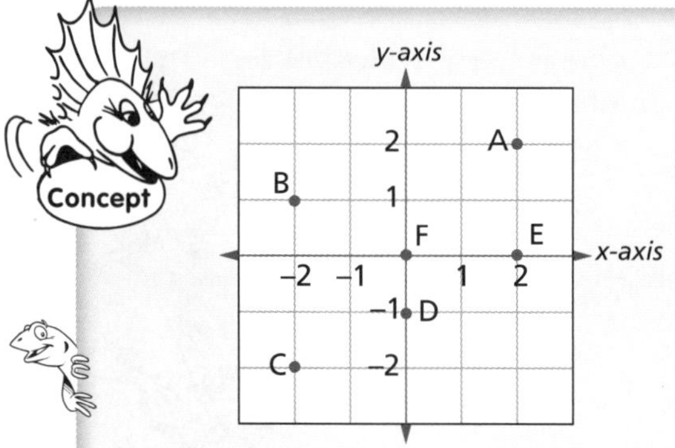


**The 4 quadrants**

Write the coordinates for:

A ____ B ____

C ____ D ____

E ____ F ____

Write the name and coordinates for a 2D shape.

____

## 37:3 ☐ out of 8

1. $\begin{array}{r} 456 \\ \times \quad 8 \\ \hline \end{array}$

2. $\begin{array}{r} 800000 \\ - \quad 7946 \\ \hline \end{array}$

3. $(5 \times 10^4) + (6 \times 10^3) + (4 \times 10^2) + (0 \times 10^1) + 4 =$ ______

4. Complete the table.

| Number of horses | 1 | 2 | 3 | 4 | 5 |
|---|---|---|---|---|---|
| Number of legs | 4 | | | | |

**a** Write a rule to describe the pattern.

______

**b** How many legs are on 592 horses? ______

5. Stickers were stuck on a dice so that the faces showed three 3s, two 5s and one 6. The dice was thrown 120 times.

**a** Which number would you expect to see most often? ______

**b** About how many times would you expect to see an even number? ______

6. **Number of siblings**

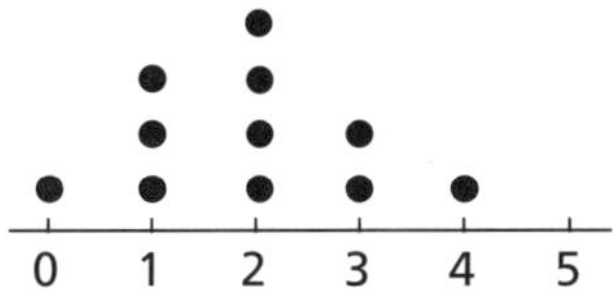


Key: ● = 10 people
(Data is rounded to the nearest 10.)

**a** How many people had 4 siblings? ______

**b** Which number of siblings was the most common? ______

7. Circle the numbers that are divisible by 3.
(The sum of the digits must be divisible by 3.)

2307 800 010 410 043 8310 7850

8. 16, 20, 24 and 28 are all divisible by ______.

## 37:4 ☐ out of 4

**Extension**

1. 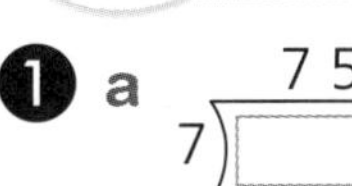
**a** $\begin{array}{r} 75 \\ 7\overline{)\square} \end{array}$ **b** $\begin{array}{r} 216 \text{ r } 5 \\ 4\overline{)\square} \end{array}$ **c** $\begin{array}{r} 11{\cdot}28 \\ 7\overline{)\square\cdot\square} \end{array}$

2. If → means 'is taller than', who is the tallest?

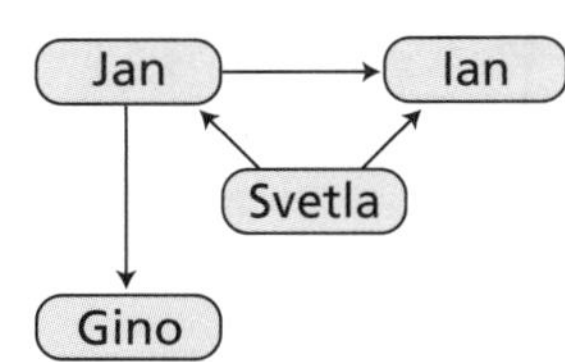


______

3. **How $2500 was spent**

| Clothing | Food | Fares |
|---|---|---|

Divided bar graph

**a** What % was spent on fares? ______

**b** What % was spent on food? ______

4. Calculate how many bird seed packets can fit in the box. ______

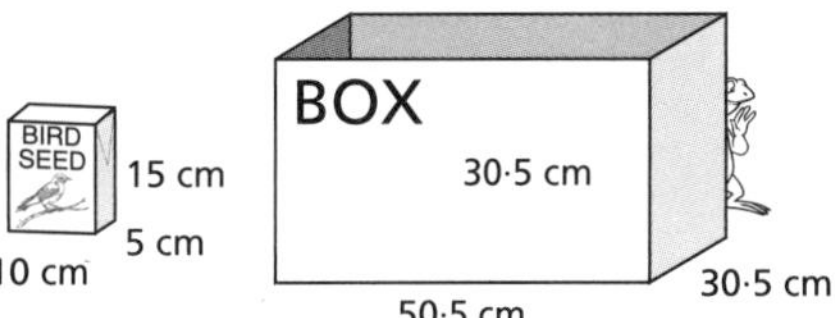


**Challenge**

*Survey 20 people. Ask them what country they would most like to visit out of Canada, Fiji, South Africa and China. Tally the results in the table.*

| | | Total |
|---|---|---|
| Canada | | |
| Fiji | | |
| South Africa | | |
| China | | |

*Discuss the results with your family.*

Fill in this table for the person measured in Unit 1.

**Name:** ______ **Date:** ______

| Age: ______ | Mass: ______ kg | Shoe size: ______ |
|---|---|---|
| Height: ______ cm | Waist: ______ cm | Neck size: ______ cm |

How have these measurements changed since Unit 1? ______

______

## Examples of measurements

1
- The width of a finger is about 1 cm.
- The length of a place-value tens block is 10 cm.

2
- The height of the girl is a little more than 1 m.

3
- The container of milk holds 2 L.
- The can of softdrink holds 375 mL.
- The teaspoon holds 5 mL.

4
- The boy has a mass of 40 kg.
- The margarine has a mass of 500 g.

5
- A double page of the newspaper has an area of half of 1 $m^2$.
- The top of a place-value ones block has an area of 1 $cm^2$.

6
- 30°C is a hot day.
- 3°C is a very cold day.

Use the pictures above to estimate the answers to these questions.

**1** a How high is the glass?
b How wide is the table?

**2** a How wide is the clothes line?
b How tall is the woman?

**3** a How much will the bucket hold?
b How much will the cup hold?

**4** a What is the mass of the dog?
b What is the mass of 2 L of milk?

**5** a What is the area of the window?
b What is the area of the top of a matchbox?

**6** a What is the temperature on a very hot day?
b What is the temperature on a cool day?

## Tables of number and measurement

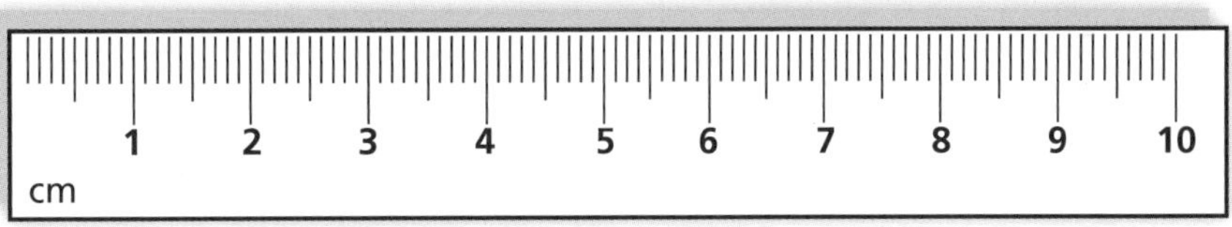


### Length

1 centimetre = 10 millimetres
1 metre = 100 centimeters
1 metre = 1000 millimetres
1 kilometre = 1000 metres

### Area

1 hectare (ha) = 10 000 $m^2$
1 square metre = 10 000 $cm^2$
1 square kilometre = 1 000 000 $m^2$
1 square kilometre = 100 ha

### Mass

1 kilogram = 1000 grams
1 tonne = 1000 kilograms
1 gram = 1000 milligrams

### Capacity and volume

1000 millilitres = 1 litre
1000 litres = 1 kilolitre
1 millilitre = 1 $cm^3$
1 litre = 1000 $cm^3$
1 litre of water has a mass of 1 kg.
1 kilolitre = 1000 litres
1 megalitre = 1000 kilolitres

A teaspoon holds about 5 mL.

5 mL

A carton of milk holds 1 L.

### Time

1 minute = 60 seconds
1 hour = 60 minutes
1 day = 24 hours
1 week = 7 days
1 fortnight = 2 weeks
1 year = 52 weeks
1 year = 365 days
1 leap year = 366 days
1 decade = 10 years
1 century = 100 years

**am** stands for **ante meridiem**.
**am** means **before midday**.

**pm** stands for **post meridiem**.
**pm** means **after midday**.

You must learn your tables!

The **freezing point** of water is **0°C**.
The **boiling point** of water is **100°C**.

°C means degrees Celsius.

A temperature of **5°C** is a **cold** day.
A temperature of **35°C** is a **hot** day.

### Roman numerals

| | | | |
|---|---|---|---|
| **1** = I | **6** = VI | **20** = XX | **90** = XC |
| **2** = II | **7** = VII | **30** = XXX | **100** = C |
| **3** = III | **8** = VIII | **40** = XL | **200** = CC |
| **4** = IV | **9** = IX | **50** = L | **500** = D |
| **5** = V | **10** = X | **60** = LX | **1000** = M |

### Months of the year

Thirty days has September, April, June and November. All the rest have thirty-one, except February alone, which has twenty-eight days clear and twenty-nine days each leap year.

You can use the knuckles of your hands to find the number of days in each month.

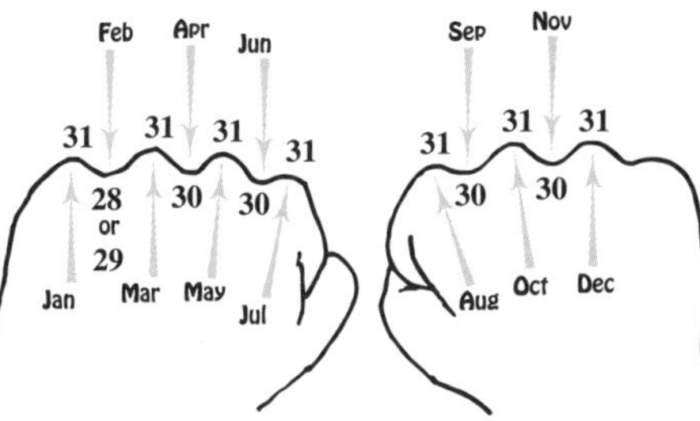


Every 4th year is a leap year.

### Seasons

**Summer:** December, January, February
**Autumn:** March, April, May
**Winter:** June, July, August
**Spring:** September, October, November

### Multiplication tables

| | | | | | | | | |
|---|---|---|---|---|---|---|---|---|
| 1 × 2 = 2 | 1 × 3 = 3 | 1 × 4 = 4 | 1 × 5 = 5 | 1 × 6 = 6 | 1 × 7 = 7 | 1 × 8 = 8 | 1 × 9 = 9 | 1 × 10 = 10 |
| 2 × 2 = 4 | 2 × 3 = 6 | 2 × 4 = 8 | 2 × 5 = 10 | 2 × 6 = 12 | 2 × 7 = 14 | 2 × 8 = 16 | 2 × 9 = 18 | 2 × 10 = 20 |
| 3 × 2 = 6 | 3 × 3 = 9 | 3 × 4 = 12 | 3 × 5 = 15 | 3 × 6 = 18 | 3 × 7 = 21 | 3 × 8 = 24 | 3 × 9 = 27 | 3 × 10 = 30 |
| 4 × 2 = 8 | 4 × 3 = 12 | 4 × 4 = 16 | 4 × 5 = 20 | 4 × 6 = 24 | 4 × 7 = 28 | 4 × 8 = 32 | 4 × 9 = 36 | 4 × 10 = 40 |
| 5 × 2 = 10 | 5 × 3 = 15 | 5 × 4 = 20 | 5 × 5 = 25 | 5 × 6 = 30 | 5 × 7 = 35 | 5 × 8 = 40 | 5 × 9 = 45 | 5 × 10 = 50 |
| 6 × 2 = 12 | 6 × 3 = 18 | 6 × 4 = 24 | 6 × 5 = 30 | 6 × 6 = 36 | 6 × 7 = 42 | 6 × 8 = 48 | 6 × 9 = 54 | 6 × 10 = 60 |
| 7 × 2 = 14 | 7 × 3 = 21 | 7 × 4 = 28 | 7 × 5 = 35 | 7 × 6 = 42 | 7 × 7 = 49 | 7 × 8 = 56 | 7 × 9 = 63 | 7 × 10 = 70 |
| 8 × 2 = 16 | 8 × 3 = 24 | 8 × 4 = 32 | 8 × 5 = 40 | 8 × 6 = 48 | 8 × 7 = 56 | 8 × 8 = 64 | 8 × 9 = 72 | 8 × 10 = 80 |
| 9 × 2 = 18 | 9 × 3 = 27 | 9 × 4 = 36 | 9 × 5 = 45 | 9 × 6 = 54 | 9 × 7 = 63 | 9 × 8 = 72 | 9 × 9 = 81 | 9 × 10 = 90 |
| 10 × 2 = 20 | 10 × 3 = 30 | 10 × 4 = 40 | 10 × 5 = 50 | 10 × 6 = 60 | 10 × 7 = 70 | 10 × 8 = 80 | 10 × 9 = 90 | 10 × 10 = 100 |